COMMUNITY GROUP HOMES

COMMUNITY GROUP HOMES

An Environmental Approach

Architecture · Research · Construction, Inc.

 VAN NOSTRAND REINHOLD COMPANY
New York

Copyright © 1985 by Van Nostrand Reinhold Company Inc.

Library of Congress Catalog Card Number 85-7565

ISBN 0-442-20832-4

Printed in the United States of America

Designed by Karin Batten

Van Nostrand Reinhold Company Inc.
115 Fifth Avenue
New York, New York 10003

Van Nostrand Reinhold Company Limited
Molly Millars Lane
Wokingham, Berkshire RG11 2PY, England

Van Nostrand Reinhold
480 La Trobe Street
Melbourne, Victoria 3000, Australia

Macmillan of Canada
Division of Canada Publishing Corporation
164 Commander Boulevard
Agincourt, Ontario M1S 3C7, Canada

16 15 14 13 12 11 10 9 8 7 6 5 4 3 2 1

Library of Congress Cataloging in Publication Data
Main entry under title:

Community group homes.

   1. Group homes. 2. Architecture—Environmental
aspects. 3. Architects and community. I. Architecture—
Research-Construction, Inc.
NA6775.C65   1985   725'.5   85-7565
ISBN 0-442-20832-4

CONTENTS

FOREWORD

The terms *deinstitutionalization, normalization,* and *community-based home* have become catchwords that convey a certain imagery yet often conceal more than they reveal about real processes and settings. There are several aspects of the perspective and contents of this book that make it a unique contribution to the general understanding of our relationship to the physical places in which we live and to the specific issue of group living. The book's subtitle, *An Environmental Approach*, reflects the reality that the physical environment is a powerful force in making our daily lives more or less livable and supportive of personal and social growth. It also reflects the reality that the physical environment alone cannot change the institutional administrative mentality that makes most community-based group homes fail in their stated goals of being dramatic alternatives to psychiatric institutions, even while life in them is intended to be of a higher quality than life in an institution. The environmental perspective provides a way of understanding the relationships among the physical environment; the people and processes that create it, change it, and manage it over time; and the limits or potential for the quality of life it can and does provide.

This environmental perspective is neither romantic nor paternalistic. It is neither abstract nor academic. It is realistic and concrete, based on the authors' own experiences in working with residents and staff in group homes. We are given a basis to analyze and critique existing or potential settings. The perspective begins with the obvious but rarely acknowledged fact that group living has problems, even when chosen and created by a group of like-minded people. Yet, in a world that has historically focused on the differences between those of us who have been institutionalized (or might be) and those of us who have not been (or probably will not be), difficulties in group homes are usually attributed to the problems of residents rather than to the places or to their management processes and procedures.

Often this definition of the problem leads to stereotypical decisions about "who can benefit" from living in a group home or to the re-creation of institutional qualities, rules, and procedures that are supposed to ensure "appropriate behavior." Whether it is a small detail—such as a note by a telephone or on a bulletin board—or a large-scale decision—such as site location or the size and variety of rooms or the management processes that determine to what extent residents have control and decision-making power—the message of this book is that an institutional environment can be created and symbolized as much in a house or apartment as in a hospital.

By focusing on the similarities among all group living situations, this environmental perspective clarifies the physical, social, and organizational issues that must be addressed in order to make such settings supportive for their residents by expanding possibilities rather than by imposing limitations. We are not presented with dogmatic directives. We are given constructive alternatives that deal with processes, not products. ARC, true to its basic belief that people should make their own decisions about their own settings, gives the reader the tools and methods, conceptual and practical, to evaluate and change existing settings or to look for or create new ones. We, as readers, benefit not only from ARC's own evaluations of the projects in which they have been involved, but from the experiences of residents and staff talking about their places, processes, and outcomes from their own points of view.

For those of us who have professional training, whether as designers, researchers, or administrators, ARC's work is a model of what can be done if we are committed to questioning our ways of thinking and working and to demystifying and sharing our skills. As a member of ARC might say, "This book promises nothing it cannot deliver." Yet it gives us much more than any book I have seen on the topic. Over the years I have learned much from the work of the ARC group. I am delighted that their work is now available to a wider audience who can benefit from their original and creative contribution to improving the quality of all of our lives.

MAXINE WOLFE, PH.D.
ASSOCIATE PROFESSOR
ENVIRONMENTAL PSYCHOLOGY DOCTORATE PROGRAM
CITY UNIVERSITY OF NEW YORK

PREFACE

Life in community group homes and neighborhood settings is the intended alternative to consigning people to institutions. But so often the dysfunctional qualities of institutional environments—the passive attitudes and feelings of noninvolvement—have been carried along, unquestioned, into the community.

This book describes precisely in what ways we see most group homes as a mimicry of institutions, rather than as their replacement. But, more important, we want this book to illustrate a positive alternative to the marginal environmental quality of group homes. Special sorts of places, shaped with a love and understanding of how people perceive space and how they use it, do exist. Only rarely, however, have we seen this concern incorporated into group homes. More often, we have seen embarrassing efforts at design and arrangement that seem to carry with them the oddest notions of how people actually live; a kind of television version of what residents either ought to appreciate or maybe just endure. Hours are spent by trained professionals in preparing "individual treatment programs" for residents, yet the environments of the homes themselves are hardly given a serious thought. We want to see this low-level conventional wisdom replaced with more deeply informed design.

We also want to illustrate an alternative to the conventional architectural processes that have produced these marginal environments. In these conventional design processes, an architect or designer (or, more often, an administrator) decides, ahead of time, what each space will be used for and just how much and what type of furniture is needed, where. Minimum building standards—no matter how inadequate—are often the major guide in this conventional process. The same prescribed way of doing things carries over to the day-to-day operation of the place. This process is the same as that used to build and operate institutions: it simply does not result in supportive places. It relegates residents to the passive role of service consumers rather than enabling them to be empowered, responsible participants.

Ironically, this conventional process then requires made-up "activities" (and the staff to sustain them) to fill residents' idle time—time made more idle by their noninvolvement in the day-to-day life of the place itself. The participatory process that we want to illustrate allows users—staff and residents both—a sense of control, ownership, and responsibility that conventional processes invariably deny.

We also want to show how space is affected by rules. Who makes the rules? How do rules affect perception of the place and perception of self?

These are difficult tasks. The difficulty is not that we are talking about some completely new building type without precedent or history. In fact, being involved in the creation of a new building type would seem more exciting. The problem for us is the opposite: what constitutes a group home is so typically ordinary and so often created by competent, caring people (who have devoted personal sweat and anguish to a cause) that it is hard to be critical in a way strong enough to force the issue of seeing community environments from a fresh point of view.

The point of this fresh view is to make supportive settings: to create the salutary environment. This is some of what we want to convey by subtitling the book "An Environmental Approach."

Finally, we hope that with this book we are creating a tool that will support the group-home movement. Every one of our criticisms carries with it a proposed solution. For, although we may seem critical and impatient, we do believe that life in most community homes is of a higher quality than it is in institutions.

The evolution of our interest in the new environmental potentials of group homes and community living parallels the deinstitutionalization movement of the last twenty years. Just as the change in focus moved from institutions to the community, so we began our work on the "back" ward of an old-line state mental hospital, moving through a series of projects, into more "open" wards, and then to community treatment and community living.

On the "back" ward, working together almost daily with staff people and the forty-five men who lived there, we spent eighteen months designing, building, and evaluating changes in the environment. One carefully measured result was a doubling in the rate of positive social interaction. While this occurred over the full course of the work, it seemed particularly enhanced by the fact that each man had a defined, private place to call his own.

This was a powerful experience, confirming our beliefs that there is more to architecture than just formalistic design and that there is tremendous potential in informed design. It also made us realize that the men we were working with acted as they did partly because of the place they were in, yet they were treated as if the problem was theirs alone and had nothing to do with the space they inhabited.

Since that early work, the process of user involvement has been the central focus of several environmental research projects designed to determine how such participation in decision-making affects the behavior of the users.

We designed and built changes in dayrooms of wards for long-term geriatric patients after a series of decision-making workshops in which half the residents participated. Significant behavioral differences were found, at each stage of the evaluation, between the residents who had participated and those who had not, even though all residents shared the same environment. Several small changes were made to the building as a result of the workshops: the addition of a snack kitchen, for instance. Some of the changes in the residents' behavior as a result of these relatively minor modifications would have been hard to predict ahead of time: the people who had participated in decision-making, for instance, spent more time awake, more time out of their bedrooms, and more time closer to others. This work in residential institutions sharpened our awareness of how places can be deeply harmful

and how easily the source of the damage can go unnoticed.

These experiences of working in institutions also changed our ways of seeing. The sorts of places that people gravitate toward and hold dear are, in Robert Sommer's words, soft rather than hard; they are personal rather than institutional, suggestive rather than pre-packaged, flexible rather than fixed, and incomplete rather than total.[1] The design process is as important as the design itself: who decides and who has power are critical to who feels ownership of the product. A good process will create settings that are supportive of the people who use them—settings that allow choice, personal control, and involvement by users; settings that also present challenge in an atmosphere of beneficial risk.

It may seem reasonable to assume that anyone spending time in an institution differs from the rest of us and has environmental needs very different from the norm. We had expected our work in institutions to enable us to point out all sorts of ways that institutional environments ought to be shaped to fit the particular diagnoses of these "different" people. But that is not what we learned. We found that, from an environmental perspective, the nature of the institution is a problem in itself. Any of us not used to institutional life would have difficulty in sharing space with many other people.

Categories and labels dramatize differences among people at the expense of minimizing their similarities. They obscure the distinction between individual and environmental limitations. Too often, architects designing for special populations seem to forget the basic qualities of human places and become preoccupied with categories and labels. The result is that human scale, individual privacy, and spatial differentiation are sacrificed. While we are not saying that individuals are without problems, we do believe that these problems—the differences—are not a valid point of departure for designing environments.

---

1. Robert Sommer, *Tight Spaces: Hard Architecture and How to Humanize It* (Englewood Cliffs, NJ: Prentice-Hall, 1974).

We do not have prescriptions for environments for these "different" people. What we do suggest is an approach that has been enriched by our becoming self-consciously immersed in the places where we have worked. The essence of this approach is that it is participatory, innovative, and creative. We do things in a gradual, step-by-step way, often weaving research questions into projects and always trying to inform design with thoughtful knowledge of how people use space. This design process can be applied to any building.

A single theme runs through all this work: places that are better designed for disabled people are better for everyone. Designing for and with people who are "different" teaches us about the needs we all share. Through empathy we gain insight.

As you read this book, you should be aware of some premises on which it is based. First is our use of the term *group home*. This phrase is intended in the most general way, to include any sort of community-based residential alternative to an institution, where a small group of people other than a nuclear family live and share space. For the most part, we have not found it important to distinguish among different types of group homes because the process we want to illustrate applies to them all. Of course, there are some differences among types that affect the course of the process, for example, the length of a resident's stay.

This book combines measured research work, empirical knowledge, and opinion. The fact that we have been deeply immersed in our work casts the "true believer" shadow on us. We have tried to write lucidly to overcome this shadow and to reason with internal logic rather than any presumed external authority. Our choices have been for the illuminating approximation over the apparently absolute; practical over academic; on-site, hands dirty, working with people, rather than from a distance.

Although the normative research—section 1 of the book—includes group homes that serve different groups of clients, the change projects—sections 4 and 5—all took place in group homes specifically designed for mental health clients.

All this work has been done in Ohio. Obviously there must be questions of whether it makes sense in other cultures and in other places. Parts of the book have been distributed, with requests for response, to many countries and throughout the United States and Canada. Though it is inconclusive to say so, the responses seem to suggest that the process we propose is appropriate to all these areas.

We hope that this book will not be viewed as a set of absolute standards to be imposed. Using it in this way would amount to seeing it as answer rather than question, product rather than process. Charles Olson has said, "The motive of reality is process, not goal."[2] We hope the book will be seen as a process model.

---

2. Charles Olson, *The Special View of History* (Berkeley: Oyez, 1970), p. 49.

ACKNOWLEDGMENTS

In addition to getting breathless over the obvious, we have also been criticized for not recognizing how much effort it takes for operators, volunteers, residents, and staff to begin a group home—any group home—and to keep it going. Also, we have had people tell us that we do not recognize how difficult it is to sustain the energy that environmental-change projects require; it is just too easy for the ordinary routine to impose itself again. This is fair criticism. We especially say thank-you to those who have been through the design process with us and still maintain a sense of why it was important; also thank-you to these same people for accepting our intrusions into their space, and most of all, for letting us use them to illustrate that environmental change is possible, even though difficult.

Many, many people contributed to the contents of this book. In listing names we are of course including some who may not endorse all we say and leaving out others who helped a great deal: Neta Berman, Jim Buskirk, Mary Campbell, Charles Curie, Tony Dattilo, Carla Davis, Gerda Freedheim, the staff of Hill House and Homewise, Cathy Hrybiniak, Toaru Ishiyama, Rosemary Kulow, Jim LaRue, Donna Lohr, Steve Morse, William Muraco, Roger Murray, Bob Obermeyer, Curry O'Connell, the staff of Panta Rhei, Dee Roth, Tina Pine, Sue Powell, Claudia Reiter, Bill Rubin, Judi Slack, Carl Strayer, Hank Tanaka, Fred Toguchi Associates, Bill Whan, Pat Warden, Toby Ann Weber, Jim Wells, David Whetsell, Charles Welch, Suzanne Welch, and Maxine Wolfe. Even with this list of names, we hope that along the way we have also said thank-you.

The A•R•C group (Architecture•Research•Construction, Inc.) is worker owned and runs consentually with as little hierarchical structure and ritualized role behavior as the formalities of being a corporation require. Michael Bakos, Richard Bozic, David Chapin, and Steven Kahn were the core group who initiated the A•R•C projects, carried them through to completion, wrote the bulk of the material, and did the illustrations; Stephanie Neuman, Ph.D., did most of the technical work in developing the research designs to gather the empirical knowledge, analyzed most of the data, and wrote about it; Judith Gandrus Burt wrote major parts of the book and participated in the projects themselves; William Mateer, as director of Panta Rhei, made his aftercare agency available as sponsor of the projects, maintained support for the work all the way through, and critiqued much of the writing; and Michael Bakos and David Chapin rewrote a great deal of the original material and edited it into this book.

Funding for research projects came from the State of Ohio, Richard F. Celeste, Governor; Department of Mental Health, Pamela S. Hyde, J.D., Director, and Timothy B. Moritz, M.D., past Director; Dr. Paul F. McAvoy, past Commissioner; through the Office of Program Evaluation and Research, Dee Roth, Chief. Research was sponsored by Panta Rhei, Inc., an aftercare rehabilitation agency, William Mateer, director and research project coordinator.

Dee Roth managed our work firmly but with supportive encouragement. She reminded us that applied research must produce useful results, yet she has never attempted to censor the questions we might ask or the answers we have produced. She has worked against incredible odds to keep a series of projects going with some continuity, stretching limited resources all the way. If there is a single hero in all of this it is Dee Roth; without her this book would not be.

MICHAEL BAKOS
DAVID CHAPIN, ARCHITECT

A STUDY OF COMMUNITY RESIDENTIAL ENVIRONMENTS

A Short History

American life has long included communal
groups. In deciding to live together,
members make a commitment to the group, often
sharing a common goal. For many this goal was
a shared philosophical or religious view—a
fundamentalist theology, a belief in the intrinsic
wickedness of the materialist world, the desire to
live in harmony with nature. Others sought
escape from an increasingly complex techno-
logical society and a freer life for children than
was possible in the outside world. Those groups
that built buildings created a legacy.

The environment—house, village, or farm—is
the setting for group life, shaped to support the
group and its goals. It became a symbol and a
reflection of the identity and personality of the
group and was often constructed to include within well-defined borders ". . . vantage points
[that] allowed members to survey the whole of
their model domain. The Shakers established
mountain sanctuaries; the Oneidans built towers;
the Mormons constructed a promenade atop their
Nauvoo Temple; the Amana Inspirationists used
a windmill as a vantage point. From here they
observed their landscapes, where the recycling of
water and waste within closed systems restated
their boundaries. They could also observe a
range of services defining their self-sufficient
economies. A typical commune would include
a blacksmith shop, a machine shop, a sawmill,
a gristmill, a printing plant, a communal store,
a school, a church, and perhaps a hotel."[1]

The rules, policies, and understandings of the
group determine how people interact with each
other and their environment. In addition to a
common purpose that brings the group together,
rituals continually reinforce the group spirit and
strengthen the sharing of work and rewards.
Issues of money, sex, and power always require
attention.

The constant tension between individual needs
and group integrity is reflected in leadership
style. Many groups depend on a charismatic,
fiery, authoritarian leader; others are egalitarian,
leaderless. Another group barometer is the decision-making process, particularly about issues of
environment. Are decisions made in a participatory way or by those in authority? How does the
need for private, individual spaces balance with
the need for communal shared space?

Successful communes were ". . . distinguished
by a number of common themes that are a legacy to the contemporary commune movement.
Members knew one another beforehand, they
invested their time and resources equally, they
owned property collectively. They had a common
faith and a common goal. They organized, shared
and rotated chores. They developed rituals to
celebrate their togetherness—celebrations, myths
and symbols, special ceremonies. They had regular meetings to make decisions, air grievances,
and work out relationships. Such practices reinforced the commitment that eased successful
communes over the inevitable difficulties of communal living."[2]

Despite the differences between group homes
and intentional communities—group homes are
planned to be an integrated part of their surroundings, not separate—the parallels are striking. Both confront the same issues of balancing
common, shared space with individual, private
space. Leadership style and personal freedom are
issues in both, just as are philosophical orientation and daily chores. It is also striking that they
both have been subjected to hostility and vio-
lence from their neighbors; especially striking
because they all share the simple desire for a
happier, more fulfilling life.

Although the basic idea of community living is
not new, group homes as a reality are. The history of community residential facilities, as an integral part of the deinstitutionalization movement,
is relatively short. In 1960, there were perhaps
as few as ten adult psychiatric community resi-
dences in the nation. Nursing homes, rest homes,
and boarding homes were the other major types
of "community" facilities available, but these
were basically "holding facilities" that provided

---

1. Delores Hayden, *Seven American Utopias: The Architecture of Communitarian Socialism, 1790–1975* (Cambridge,
MA: The MIT Press, 1976), p. 43.

2. Rosabeth Moss Kanter, "Communes for All Reasons," *Ms*
(August 1974), pp. 62–67.

long-term care for people who had been confined to institutions but were not primarily in need of mental health care. In 1963, the federal government passed the Mental Retardation Facilities and Community Mental Health Centers Act (Public Law 88-264), which laid the groundwork but did not actually provide funding for community residential facilities. Its focus was short-term, crisis-oriented prevention. (The continuation of this law, the Health Revenue Sharing Act of 1974, did include provisions for funding of residential facilities.) In 1968, to provide treatment for people convicted of (usually nonfelonious) crimes and of drug abuse, Congress passed the Omnibus Crime and Safe Streets Act, which funded various types of halfway houses.

A 1974 survey was able to locate 289 community-based psychiatric facilities, more than double the 128 counted in 1969.[3] By 1985 the rapid rate of increase made the total difficult to know, but a reasonable estimate is 10,000 facilities. Adding other groups of vulnerable people living in community facilities—elderly, mentally retarded, and people within the criminal justice system—would double this total to approximately 20,000. This increase reflects the shift in emphasis from custodial care in large state institutions to rehabilitation in community-based facilities.

Several factors are generally acknowledged as giving momentum to the movement toward community care. The introduction of psychotropic drugs in the 1950s and their general use in the 1960s enabled many people to function again in the community. The 1950s also saw a reaction, which continues today, against the general ineffectiveness of large institutions in dealing with the individual's problems; the "therapeutic community" was voiced as an alternative. More recently, institutions have become embroiled in "right-to-treatment" suits in the federal courts. Finally, the high cost involved in operating large institutions has made these facilities unpopular with taxpayers and politicians alike.

Although the concept of the therapeutic community was originally developed in a hospital setting, it provides the basis for many current rehabilitation efforts in group homes. All of the events of day-to-day life are seen as part of the treatment or rehabilitation effort, with ". . . the treatment agency [becoming] the total environment and the total environment [being] the agency for treatment. Activities from morning till night [are] to help the patients prepare for community living."[4] Often, the greatest need is for practice in dealing with the events of everyday life: the seriousness of the need is proportional, usually, to the length of institutional confinement.

In 1955, there were 559,000 people in public mental institutions in the United States; by 1975, that number had declined to approximately 200,000, and to 133,550 by 1980.[5] These often-cited statistics give the impression that deinstitutionalization is a fact in the United States. The same statistics are sometimes given as an explanation of the apparently increasing numbers of homeless people, and, thereby, turned into an argument against all community care. There are, unfortunately, some realities buried by omission in these figures: the actual number of institutionalized people has increased—not decreased—significantly over the past twenty years. More mentally disabled people—750,000 in 1980—are now residents of nursing homes than were ever housed in state hospitals.[6] There are also more people in private psychiatric hospitals and in psychiatric wings of general hospitals.[7]

Rather than a result of deinstitutionalization, homelessness appears to be a problem linked to

---

3. Richard D. Budson, *The Psychiatric Halfway House* (Pittsburgh: University of Pittsburgh Press, 1978), p. 10.

---

4. Maxwell Jones, *The Therapeutic Community* (New York: Basic Books, 1953), p. 12.

---

5. *Statistical Abstract of the United States: 1984* (Washington, D.C.: U.S. Government Printing Office, 1984), p. 120.

---

6. For this and other statistics describing the location of mentally disabled people, see: U.S., Department of Health and Human Services, *Toward a National Plan for the Chronically Mentally Ill* (Washington, D.C.: U.S. Government Printing Office, 1980).

---

7. For an excellent analysis of the changing locale of care, as well as an in-depth study of services that community residents receive, see: Solomon, Phyllis; Gordon, B.; and Davis, J.; *Community Services to Discharged Psychiatric Patients* (Springfield, IL: Charles C. Thomas, 1984).

two trends: the increasing polarization of incomes toward more rich and more poor, with fewer middle class; and the ongoing destruction of the low-cost housing stock. In other words, homelessness is more likely a result of poverty than of a failure of community care.[8]

In the past few years, however, there is no question that resistance to the group home movement has developed on many fronts: among local citizens and neighbors concerned about quality of life, political and civic leaders seizing a volatile issue, and professionals concerned with protecting their bailiwick. Despite studies to the contrary, home owners worry about falling property values and neighbors fear increased crime caused by the presence of a group home.[9]

There have been charges that deinstitutionalization has merely meant a large-scale "dumping" of patients who are unprepared for life outside the institution into communities that are unprepared to meet their needs. This dumping is particularly tied to "welfare hotels" and other mini-institutions that are not at all within the spirit of the deinstitutionalization movement. Media exposés and serious editorials add fuel to this controversy, which, in some cases, is justified, for there are instances of neglect, outright abuse, and profiteering connected with some group homes.

All in all, the emerging picture is of an often fearful society not nearly as tolerant of deviancy as the figures cited earlier might indicate. These figures also fail to point out that many of the earliest group homes were established in run-down, low income areas and so were not perceived as a real threat to more affluent neighborhoods. Now, as group homes are being established in better areas, old boundaries are threatened. This may account for some of the increasing resistance to group homes. This fact couples ominously with attempts to dismantle various federal human service programs.

Unfortunately, most of the public debate over community facilities has occurred in emotionally charged, adversarial hearings on proposed changes in local zoning regulations. The outcome is often a push for more stringent standards.

New government standards might serve to improve the quality of community facilities and increase their numbers. Just as likely, however, new standards will make group homes more like the institutions they were intended to replace. Requiring that all group homes install commercial sprinkler systems would have this effect.

The group-home movement has been sustained largely by a mixed group of people—operators, professionals, and family support groups—willing to take a risk. Usually they possess somewhat maverick, pioneering energy. Forcing group homes to meet unnecessarily rigid standards might only make them more "professionalized" and, in the process, drive some of these energetic people away without offering equally competent or equally spirited replacements. Raymond Glasscote, in the foreword to his book on halfway houses, puts it this way: "Indeed, one might

---

8. A recent study of nearly 1,000 homeless people found that more than two-thirds of those interviewed had never been hospitalized in any sort of psychiatric hospital, public or private. The majority of responses to "Reported Major Reason for Homelessness" were economic. See: Ohio, Department of Mental Health, *Homelessness in Ohio: A Study of People in Need*, NIMH #1R18 MH38877-01, prepared by D. Roth, et al. (Columbus, OH: The Ohio Department of Mental Health, February, 1985), p. 36 and p. 65.

9. See the following:

Rabkin, Judith G., Muhlin, Gregory; and Cohen, Patricia W., "What the Neighbors Think: Community Attitudes Toward Local Psychiatric Facilities," *Community Mental Health Journal* 20, no. 4, (Winter 1984).

Ohio, Metropolitan Human Services Commission, *The Non-Effect of Group Homes on Neighboring Residential Property Values in Franklin County*, prepared by Christopher A. Wagner and Christine M. Mitchell (Columbus, OH: Metropolitan Human Services Commission, August 1979).

New York, Office of Mental Retardation and Developmental Disabilities, *Group Homes for the Mentally Retarded: An Investigation of Neighborhood Property Impacts*, prepared by Dr. Julian Wolpert, Albany, NY: New York State Office of Mental Retardation and Developmental Disabilities, August 1978).

Dear, Michael, "Impact of Mental Health Facilities on Property Values." *Community Mental Health Journal* 13, no. 2, (1977).

Lansing, Michigan, City of Lansing Planning Department, *The Influence of Halfway Houses and Foster Care Facilities Upon Property Values*, 1976.

Macon County, Illinois, Macon County Community Mental Health Board, *The Impact of Residential Care Facilities in Decatur*, prepared by Zack Caulkins, John Noak, and Bobby Wilkerson, (Decatur, IL: Macon County Community Mental Health Board, December 1976).

conclude that the greatest advantage that the halfway house movement has enjoyed during its forward surge of the 1960's had been precisely the lack of official guidelines.''[10]

Notes on Terms

For the most part we have tried to use words in their ordinary sense, but there are some words that we use to convey a particular meaning.

Group Home. A group home in this book refers to *any* community living arrangement for people who might otherwise be consigned to institutions. It includes lodges, apartments, halfway houses, community residential facilities, and sponsor-care homes. On the other hand we do not include welfare hotels and other mini-institutions of over fifteen residents in our definition.

Group home does not refer to one particular approach to running a community residential facility; rather, it includes any community place that is intended to be a part of the deinstitutionalization movement.

Institutional. Institutional in this book connotes a style, a mind set, a way of thinking—not necessarily a building. Institutional thinking embraces roles, rules, categories, and labels that are rigid rather than flexible. The perpetuation of the place seems more important than individual growth and development.

Institutional thinking does not occur only in institutions, it invades group homes as well. Institutional group homes perpetuate passivity and dependence by continuing to treat residents as a mass and not as individuals, by continuing the tradition in which rules and restrictions are made by staff and imposed on residents, and by continuing humiliating procedures that limit personal freedoms, choice, and options. In group homes, institutional thinking is a style to be avoided.

Supportive Settings. Perhaps the best definitions of the term supportive settings as used in this book can be found in the following quotations:

A behavior setting is a place where most of the inhabitants can satisfy a number of personal motives, where they can achieve multiple satisfactions. In other words, a behavior setting contains opportunities.[11]

A place is a configuration of possibilities.[12]

[Supportive places provide] conditions and opportunities that help people realize their full potentialities as alive human beings; that is, places that help people accumulate and assimilate experience which contributes to their sense of self worth, their sense of self capability, and their sense of life meaning; places that help people to be whole, confident, and happy.[13]

Behavior settings and their typical physical settings are a sort of cultural gyroscope. They stabilize and define a particular culture by transmitting familiar expectations and eliciting familiar patterns of behavior. A familiar physical setting, therefore, calls forth a predictable behavioral pattern that causes a minimum of overt stress.

It would be hell if every situation were new to us and we did not know how to interpret our setting, or the behavior of others. That is, in fact, the particular hell experienced by some upon entering a new culture. It is the hell of a person who is not ''adjusted'' to society and its expectations—a person whose perceptions and responses are not in the ''script.'' It is a hell that we often label insanity.[14]

---

10. Raymond M. Glasscote, et al., *Halfway Houses for the Mentally Ill* (Washington, D.C.: Joint Information Service Publication, 1971) p. x.

11. Roger G. Barker, ''Ecology and Motivation,'' in *Nebraska Symposium on Motivation*, edited by Marshall R. Jones (Lincoln: University of Nebraska Press, 1960), p. 25.

12. Ohio, Department of Mental Health, *Handbook: Changing Places and Settings*, prepared by Architecture•Research• Construction, Inc., 1975.

13. William Kleinsasser, ''Experiential Design Considerations'' (manuscript, University of Oregon, School of Architecture, 1974), p. 1.

14. Sym van der Ryn, ''The Behavior Setting: Module for Design,'' *Environment/Planning and Design* 8 (July–August 1970), pp. 14–16.

Normalization. As clumsy as this word is, it still most closely represents the chosen direction of several current social movements: deinstitutionalization, mainstreaming, and desegregation. Normalization means keeping ordinary life, within each person's abilities, as intact as possible rather than allowing disabilities or differences to dictate an insular or strange way of living.

Normalization signifies the positive value gained by each individual in the sharing of everyday life—to be part of the mainstream. There is also positive value gained by different people associating with each other, experiencing directly both their sameness and their diversity. The whole fabric of democratic society is strengthened by all its members associating and interacting. It is really a spectacular concept.

There is a term associated with group homes that we find extremely disturbing. The term is *homelike*, as in "We're proud of our homelike atmosphere here." To be homelike means to be almost but not quite a home: to mimic the look of a home rather to achieve the dynamic, lively (sometimes cluttered) qualities of a real home. It results from a failure to look beyond surface appearances to the essence of what a home is. It may be profitable for chain motels to use an artificial homelike atmosphere to sell homes away from home, but group homes should do better than motels.

Environment. A dictionary definition of environment is: "the external circumstances, conditions, and things that affect the existence and development of an individual, organism, or group." We often are referring specifically to the physical environment, which includes the process of managing and controlling the physical environment. We are using "environment" or "physical environment" to refer both to a place and to its rules, not to just buildings alone.

Institutions versus Group Homes

At their best—around the turn of the century—total residential institutions were the progressive alternative to incompetent family care or outright maltreatment. They provided a place of refuge large enough for trained employees to provide hygienic care economically. The buildings themselves were a source of great civic pride. Even today, some of the remaining originals retain an impressive grandeur and an awesome sense of purpose.

By the 1950s institutions had devolved into gigantic human warehouses. Deinstitutionalization began as a reaction to this warehousing and was facilitated with the use of psychotropic drugs in the 1960s. The community movement has focused on the failure of the institutional approach.

Habits and behaviors appropriate to institutions hinder reintegration into the natural community. Understandably, an institution establishes its own internally defined routines based largely on patterns of staffing, shift changes, and budget allocations. It is, by definition, a closed, restrictive environment with its own rules and standards. An inmate once adapted to institutional life may find it difficult—sometimes impossible—to become reintegrated into ordinary life, to function again productively, or to shake off the stigma of having been institutionalized.

Anyone who spends time in any institution, be it prison, the military, or a mental hospital, acquires behavioral patterns that may or may not be useful in the larger society. In the military it may be important to fold clothes in a prescribed way, to follow orders unquestioningly, and to show no initiative. In prison survival may depend on alternating between submissive and aggressive behavior. Patients in a mental hospital may learn to go to bed at nine o'clock and never to argue with anyone in authority.

Community living means not taking on such maladaptive institutional behaviors. Daily contact with the people and routines of the community can raise the expectations and actual abilities of the residents of group homes. Models for better functioning are readily available. Regular contact with people in the community also provides contact with the standards, mores, and behaviors of the general public.

Since the group home is located within the larger community, the residents can make use of local services such as doctors, dentists, amusements. To get them, residents must act; they must get out there. Going to a local doctor or dentist, signing up for a night class at the public school, attending church, seeing a movie, all provide opportunities to interact with people along the way—the owner of the coffee shop, the bus driver, the noisy kids at the bus stop.[15]

Institutions foster a lifestyle of passivity and dependence. Meals are served, beds are changed, and clothes are washed in the central laundry. The activities left to residents serve only to pass endless time: watching television or, at best, making ceramic ashtrays in the activity therapy program.

Without active participation in the day-to-day business of living, residents lose simple basic skills—planning a meal, making choices, shopping, cooking, cleaning, washing, and so on. The diverse cause-and-effect relationships of everyday life are replaced by authoritarian rules and conformity. Lost skills also make it difficult to reintegrate into the outside world.

In group homes, the skills of daily living are not abandoned. Everyday problems and stresses are opportunities to acquire and maintain survival skills. By preparing clothes for the morning, catching a bus, expressing a gripe on the job, paying the rent, choosing to do the ironing or to snooze, residents face the same realities that everyone in the community faces. In the most effective group homes, ordinary and functional behaviors are necessary to make it through the day. The routines and rhythms of the group home are in step with those of the community fabric.

Institutions treat people as a mass, rather than as individuals. In institutions, quantity equals economy and efficiency, and therefore thirty or forty or more residents are grouped together. For life to function smoothly, the group must be dealt with as a whole. Everyone must conform, regardless of the loss of self esteem. Little choice is left to the resident about schedules— whether to shower on Monday or Wednesday, get up late in the morning, play bingo, or work

on a hobby. Residents rarely leave the grounds alone for shopping, social activities, or productive work; trips are always taken as a herd.

The pervasive message in institutional spaces is antipersonal; the building and many of the rules governing the use of space say, in effect, "You will pass through, change, but leave no mark on me. I will still be here after you have gone."

Almost by definition, institutional space belongs to no one. This lack of sense of ownership in both private and state-run institutions stultifies even the simplest environmental initiative. Business administrators are forever asking themselves why nobody will close windows on cold days or turn off lights in empty rooms.

Vandalism is constant despite the fact that practically everything not vandal-proof has already been removed. Some residents may be motivated by their desire to be "someone," to make a mark. Living in a barren, impersonal space for which they feel no responsibility, they, not surprisingly, feel passive, angry, or burned out. They may feel adrift in space, living with inadequate territorial definition. It is very disorienting.

Group homes are small enough to treat residents as individuals. Ideally, in group homes, routines and supports are thoughtfully planned and continually revised. To each of the stresses of daily life, the staff response (if required at all) is tailored to the level of each person's readiness to manage the problems alone. Individuals benefit simply by being in a small group. There can be deeper interpersonal contact and more positive mutual support than in a large group. The small group often develops its own standards and goals, with house decisions made in regular meetings. Residents share responsibilities, have feelings of being a real part of something, and need not be treated as underlings.

Despite court orders, media exposés, and many attempts at reform, institutions create an atmosphere in which unmonitorable abuse flourishes. With their huge budgets and hundreds of employees, institutions become power bases for bureaucrats, inviting manipulation and deviousness. Employees become perpetuators of the system itself and help maintain the status quo. Institutions offer job security as a reward at the

---

15. For an insightful view of life in the community, see: Estroff, Sue E., *Making it Crazy: An Ethnography of Psychiatric Clients in an American Community* (Berkeley: University of California Press, 1981).

expense of initiative. Responsibility is invariably separated from authority so that staff burnout is a job hazard. On average it seems to take about two years to wear down even the most progressive or enthusiastic employee. Our friend Bill Whan put it this way:

. . . the shame, the abuse, the insulting stupidity of what goes on, the outrage of human storage called hospitalization, the dehumanization which eventually overcomes and inures us all, is not a defect in an otherwise benign system—it is the central and determining characteristic of an inherently evil system, a system which victimizes every single participant. This is the single most frustrating fact—that no matter how unthinkable conditions or incidents may be, there are no perpetrators, only victims. Every despot has his own despot, and his own set of paternalistically benign pretexts.[16]

Possibilities for exploitation and abuse do exist in group homes, both from within the home itself and from the outside community. The question in group living is how best to deal with abuse when it does occur. Sometimes the strength of the residents as a group is used to deal with neighborhood toughs or unscrupulous shopkeepers; peer support groups of residents under the guidance of a trained professional have been formed to counteract sexual exploitation and rape. Self-governance has sometimes checked unfair practices by operators. Residents do, however, remain vulnerable, even in the best facilities.

The existence of institutions perpetuates the idea that as a society we do not share responsibility for others' problems. Society seems to be saying send them off, let the "experts" deal with them. The institutions themselves have come to be seen as spooky, foreign places of dark imagery and to be labeled everything from nut house and looney bin to old folks' home and Happy Haven.

---

16. Ohio, Department of Mental Health, *Handbook: Changing Places and Settings*, prepared by Architecture•Research•Construction, Inc., 1975, p. 9.

Group homes, on the other hand, do not conjure up such negative images. They are too ordinary. Unlike the institutions, which are, by and large, modeled after hospitals sheltering people in order to treat or "cure" their sickness, group homes are modeled after the family home and concentrate on developing people's assets and abilities within the daily routine. By their integration into the fabric of society, group homes deliver a message very different from the institutions: We are all, each of us, individually susceptible to being damaged; and collectively, as members of the human family, we can share with experts the process of mending and repair.

The activities, events, and interactions of everyday life are what living in a community group home is about. Making a cup of coffee, cleaning a messy room, laughing with a friend, washing out a pair of shorts, falling down before the finish line, playing the harmonica are events and happenstance that together make up everyday life. Coping with joys and frustrations, excitement and boredom, is the stuff of ordinary life.

Group homes are an alternative to confinement in institutions. Too often though, a community group home is thought of as no more than a sheltered place to live during rehabilitation—therapy at separate, specialized centers. But why should everyday life be separated from the rehabilitation process? Sharing in both the responsibilities and the satisfactions of everyday life is essential to living in the larger, natural community.

A measure of normalization is the extent to which residents of a group home are allowed to practice necessary life skills in a context of respect and dignity. The approach to problems of everyday life provides the critical point around which the character of a group home, normal or institutional, develops.

The design, physical layout, furnishings, and use of space cannot alone make programs function successfully. But the place can work in concert with programs of normalization, or it can hinder those efforts. The physical environment can offer choice, stimulate interest, and invite activity.

To understand how normalization programs and supportive settings can be woven together in real situations, we surveyed a broad range of group living arrangements as they now exist, and we used many research tools, each with a different point of view, to conduct this survey. To compare the homes, we used the concept of a continuum and ranked the homes according to the degree to which they encouraged normalization.

Mental health as well as other types of facilities were surveyed. We visited fifteen homes many times over the course of the survey, and seventy-one other homes responded to lengthy questionnaires. The sites visited included community-based psychiatric group homes, an institution-based small group residential program, a group home for mentally retarded adults, a halfway house for prisoners, sponsor care homes, a large nuclear family, a women's collective, and a small vegetarian commune. Our aim was to cover a broad spectrum, from a supportive place such as the family home to a more rigidly run place such as the institution-based home or halfway house.

Each site visit involved interviews with both staff and residents, the use of checklists to examine the building, and behavioral mapping to document the use of space and levels of activity and interaction. In each case, the type of program operating in the home was carefully considered.

Certain problems were common to various group living situations. Conflicts arise when people must share space, as well as when they seek privacy. Schedules, activities, routines, roles, meals, responsibilities, and cleaning are some potential sources of conflict between personal and group needs. In the houses surveyed, many approaches were used to solve these problems.

Few good models of supportive group living arrangements exist. The military, prisons, and mental hospitals are certainly not. Fraternities and sororities perhaps come closer. The intentional community groups of the nineteenth century or of the 1960s are possibilities. The family has often been suggested, but there are limitations to this model, for a group of unrelated adults with similar needs is different from a family where needs vary with each individual's age. Moreover, the nuclear family model (rather than an extended family model), with its often stress-

ful role and economic requirements, is sometimes cited as a cause of the individual's problems rather than their solution.

It is important to recognize how difficult it is to operate an effective group home. Aside from the lack of good models, there are problems with budgets, inflexibility, paperwork, community attitudes, staff, and residents—all conspire to make life difficult. It is no wonder that the stuff of ordinary life sometimes gets lost. Because some group homes provide residents greater opportunities for normalization than others do, they serve as good examples, but all have room to grow.

Methods and Approach

In making this study of group homes, we used a wide variety of research tools. Our intent was to give as broad a picture as possible, from many different vantage points, and not to approach the subject from a narrow perspective.

We began with several assumptions about group living. For example, when a group of people share a house, there are common problem situations to be resolved, such as daily routines, responsibilities, and activities, the sharing of space, and the finding of privacy. It was also assumed that this is true for any group of people and not just for those in identified populations.

We also assumed that the different processes used to resolve these problems in each group-living situation could be ranked on a continuum that ranged from institutional to normalized. Our third assumption was that the homes selected to participate in this study were representative of group living, and therefore, the results of this study could be generalized to homes in other areas.

These assumptions were explored by analyzing the people who compose the group, the place or physical environment, and the program or approach for dealing with group and individual needs.

Research teams, composed of one project

member and two professional interviewers, made direct on-site visits of selected group homes in the northeastern Ohio area. The interviewers, who had no knowledge of the initial assumptions, questioned both staff and residents, and observed the activities, interactions, and use of space occurring during the evening meal at each house. Additionally, an architectural assessment of each house was made, which included photographs of the exterior and interior, sketches of floor plans, and notes about the structure and layout. Finally, a narrative description of general characteristics and subjective impressions was written.

Of the fifteen group homes selected for on-site visits, eight serve psychiatric populations. The remaining seven are nonpsychiatric facilities servicing both identified (alcoholic, retarded, delinquent, forensic) and nonidentified, voluntary (family and communal group) populations. Capacity ranges from five to twenty-two residents, but at the time of our visits, the number of residents living in these settings ranged from four to eighteen. Of the total possible sample of 139 residents, 97 actually participated in the study. This was due to low census in a group home, a refusal to participate in the voluntary study, or inability to tolerate the demands of the interview. The age of the residents ranged from twelve to seventy years, with a mean of 31. One staff member at each house who had direct, daily contact with residents was selected for the individual staff interview.

PASS 3

To assess the overall quality of each program based upon the principles of normalization, we used the *Program Analysis of Service Systems: A Method for the Quantitative Evaluation of Human Services*, known as PASS. This method was developed by Wolf Wolfensberger and Linda Glenn with the support of the National Institute of Mental Retardation of Canada and was published in 1975. It was developed initially in response to the growing need for an assessment of the quality of human service delivery systems. As Wolfensberger notes in his introductory com-

ments to PASS 3, ". . . much of what is practiced in human management services rests on quite deeply ingrained dehumanizing attitudes and values . . . particularly when these services are directed toward populations of people who are considered deviant or devalued." [17]

PASS enables the researcher to explore carefully the means and processes of programs in achieving their stated objectives. These universal human service principles are based upon the ". . . use of means which are as culturally normative as possible in order to elicit and/or maintain in potentially 'deviant' clients, behaviors which are as culturally normative as possible." [18]

PASS 3, the third edition of the study, includes fifty major characteristics, such as accessibility or environmental beauty, regarding the quality of the program. Each of these characteristics is rated on a scale of objective and concrete levels.

To determine the level ranking, several factors, varying from one characteristic to another, are scored on a scale from very poor to optimal. For each characteristic measured, the lowest level reflects unacceptable performance. The total PASS 3 score is the sum of all the rating scores based upon the level weights and may range from minus-947 to plus-1000. The top figure of plus-1000 represents the ideal program that fully supports the principles of normalization.

For our study, a short form of PASS 3 was developed covering fifteen major problems in community-based group living. A conversion formula was used to compute the total score for this modified short version. (More complete information and descriptions are available in the PASS Manual and PASS Handbook. National Institute on Mental Retardation, York University Campus, 4700 Keele Street, Downsview, Toronto, Canada.)

17. Wolf Wolfensberger and Linda Glenn, *PASS 3: Program Analysis of Service Systems* (Toronto: National Institute of Mental Retardation, 1975), p. 6.

18. Wolf Wolfensberger, *The Principle of Normalization in Human Services* (Toronto: National Institute of Mental Retardation, 1972), p. 28.

Normalized-to-institutional Continuum

The total PASS score derived from the modified version of PASS 3 was used to assess the degree of normalization versus institutionalization in each of the group homes. The range of these scores paralleled quite closely those that have been found in previous research; therefore, logical statistical groupings were possible. The group homes were categorized in four primary divisions based upon the mean and standard deviation of the PASS 3 score. At one extreme of the continuum were three "institutional" group homes, whose total PASS 3 scores placed them below minimally adequate standards for group living. Next along the continuum are the categories designated as "Low-Medium" and "High-Medium" (each having four homes) whose scores indicated that they had met minimum expected standards but had not achieved an expected level of performance for a normally supportive environment. Finally, four group homes had PASS 3 scores that were considered to be within the normally supportive range (between 712 and 999) and surpassed expected performance levels but were not ideal.

Our version of PASS 3 addressed a number of questions about the quality of group home environments. Some items from the original PASS 3 that we selected for use in our modified version are listed below.

● Are physical resources such as banks, grocery stores, clothing stores, movie theaters, parks, and other recreational facilities within walking distance or easily accessible by public transportation?
● Does the home blend into the surrounding neighborhood or stand out from it? Are there signs that draw unusual attention to the building? Are unnecessary safety features, such as heavy mesh screens and alarms evident on the exterior, making the house stand out?
● Does the role expectation for the clients support deviancy and further "dehabilitation" by depriving them of normal social contacts? Are most daily contacts within the group or with

TABLE 1-1. RESULTS AND STATISTICAL INTERPRETATION OF PASS 3

PASS 3 CONTINUUM

−947 to 0 = Unacceptable program quality
0 = Minimally adequate level of performance
+711 = Expected level of performance
+1000 = ideal program fully supporting principles of normalization

STUDY RANGE OF PASS 3 TOTAL SCORES

Range = −50 to +980
Mean + 400.875
Median + 411.500
Standard Error = 87.323
Standard Deviation = 349.291
±1 standard deviation = +60.584 to 750.166

CONTINUUM RANKING RESULTING FROM THIS STUDY

Institutional	Low Medium	High Medium	Normally Supportive
−50	+137	+424	+738
−38	+207	+460	+886
−11	+305	+497	+916
	+406	+557	+980

the broad spectrum of society? Are contacts with the general public encouraged through work, recreation, and worship?
● Are outings always in large groups, or can residents become involved individually with outside activities?
● Are activities age-appropriate and valued within the broader context of the community and society?
● Are furnishings and decorations appropriate?
● Is attention paid to the residents' personal appearance? Is appearance appropriate to the residents' age and culture? Or do residents look different from the general public, easily identifiable as "institutionalized"?
● Are routines of retiring, awakening, eating, and the like culturally normative, or are they defined by agency or staff schedules?
● Is there a good balance between active and passive leisure time? Is there too little to do?
● Would the typical day be considered normal for the large society?

- Are normal built-in risks eliminated? Does the place unnecessarily limit exposure to normal dangers, risks, and growth challenges?
- Is furniture heavy duty or institutional in appearance? Are there special windows and screens? Is furniture chosen for appearance rather than comfort? Is the place too hot or too cold? Is it excessively noisy?
- Is the place drab, barren, and monotonous in appearance? Have there been superficial attempts at beautification? Is there a variety of lighting?
- Is there adequate privacy in bedrooms and bathrooms? Are there choices of single and multiple bedrooms? Can residents decorate their bedrooms or personal spaces?
- Are client-staff relationships open and direct? Are there obvious attempts to differentiate between roles? Do staff members knock on clients' bedroom doors before entering?
- Are interactions between clients appropriate and adaptive? Is this encouraged by the staff?
- Are there a range of settings for a range of group sizes?

Resident Interviews

Each resident who agreed to participate in the study was individually interviewed. Several measures were selected to better understand how people living in community group settings felt about their current situations as well as to explore the kinds of activities they were involved in on a day-to-day basis.

During the course of the interview, the resident's social history was taken using a modified version of an extensive questionnaire (table 1-2).[19] Another questionnaire (table 1-3) explored the resident's degree of participation in such free-time activities of daily living as going to a movie, watching television, visiting, and reading. It was drawn from the Katz Adjustment Scale.[20]

TABLE 1-2. SOCIAL HISTORY INVENTORY: STATISTICAL SUMMARY OF TOTAL SURVEY RESPONSE

	Mean	Range
Resident age	31	12 to 70
Marital status	78% single	
Education	10-11th grade	
Income	$3000	$0 to $20,000
Length of residence at present address	< one year	
Prior living situation	hospital	
Next living situation anticipated	alone	

FREQUENCY OF RESIDENTS' RESPONSES TO INCOME LEVEL

1. No income	********************************* (33)
2. under $3000	************************************* (37)
3. under $5000	************ (12)
4. $5-10,000	********* (9)
5. $10-20,000	** (2)
6. over $20,000	* (1)

A third questionnaire determined how the residents felt about their current group-home setting and situation (table 1-4). We used a modified version of a questionnaire that had previously been used in community group-living research with adult retarded populations.[21]

Rotter's Internal/External Scale was also given to each resident, either as a paper-and-pencil task or read aloud by the interviewer.[22] This measure explores the degree to which people believe their life to be controlled by them (internal locus of control) or by society and others (external locus of control). Reliability for this scale has been quite adequate, as well as its application to a wide variety of populations.

In our study the Internal/External Scale did

19. Benjamin Pasamanick, et al., *Schizophrenics in the Community* (New York: Appleton-Century-Crofts, 1967).

20. M. M. Katz and S. B. Lyerly, "Methods for Measuring Adjustment and Social Behavior in the Community: I. Rationale, Description, Discriminative Validity, and Scale Development," *Psychology Reports* 13 (1963), pp. 503–35.

21. F. C. Scheerenberger and D. Felsenthal, "Community Settings for MR Persons: Satisfaction and Activities," *Mental Retardation* 15, no. 4 (1977) pp. 3–7.

22. J. B. Rotter, "Generalized Expectancies for Internal Versus External Control of Reinforcement," *Psychological Monographs* 80, no. 609 (1966), p. 46.

TABLE 1-3. LEVEL OF FREE-TIME ACTIVITIES: STATISTICAL SUMMARY

Activity	Total Survey % Response	Breakdown of the Four Continuum Groups				p**
		Institutional Group % Response	Low Medium Group % Response	High Medium Group % Response	Normally Supportive Group % Response	
Help with chores	75–F	68–F	81–F	69–F	81–F	*
Visit friends	42–N	68–N	42–F	46–N	40–F	*
Visit relatives	49–N	74–N	36–S&N	50–N	45–S&N	*
Entertain friends	64–N	84–N	61–N	65–N	48–N	*
Household budget	78–N	89–N	80–N	64–N	81–N	*
Engage in social activities	43–N	75–N	59–N	54–N	40–F	p < .03
Get along with neighbors	58–N	61–F	65–N	59–N	75–N	p < .01
Keep bedroom neat	86–F	100–F	75–F	91–F	87–F	*
Exercise	43–F	47–S	45–N	56–F	52–F	p < .03
Listen to radio or records	64–F	55–F	58–F	68–F	69–F	*
Watch TV	49–F	63–F	65–F	46–F	38–N	p < .01
Attend community meetings	68–N	90–N	74–N	76–N	43–F	p < .001

Responses to questions were: Frequently (F), Sometimes (S), Never (N). Responses are reported in percentages and modal response.

* Nonsignificant difference between groups; p is significant only when less than .05.
** Significance levels were determined from chi square analysis of the four continuum groups. Value p denotes the odds of a particular result being found by chance.

TABLE 1-4. RESIDENT SATISFACTION QUESTIONNAIRE: STATISTICAL SUMMARY

	Total Survey % Response	Breakdown of the Four Continuum Groups				p**
		Institutional Group % Response	Low Medium Group % Response	High Medium Group % Response	Normally Supportive Group % Response	
Do you like living here?	92 yes	90	83	100	100	*
Would you like to live elsewhere?	58 yes	44	74	52	58	*
Do you have a special friend?	65 yes	37	70	82	69	p < .02
Does that person live here?	68 no	75	80	50	67	*
Do you like the food here?	86 yes	90	65	100	92	p < .006
Do you get enough to eat?	92 yes	100	83	91	100	*
Can you decorate around your bed?	86 yes	68	83	95	100	p < .04
Can you go outside when you want?	90 yes	95	91	100	62	p < .003
Can you pick out your own clothes?	97 yes	100	96	100	92	*
Do you ever have money to spend?	90 yes	90	91	91	85	*
Do you have a special place for things?	96 yes	100	96	96	92	*
Are they safe?	88 yes	95	96	76	85	*
Do you get a good night's sleep?	90 yes	79	91	91	100	*
Do you feel comfortable here?	95 yes	94	87	100	100	*
Do you have a special place to sit?	64 yes	74	44	68	77	*
Do you like to be with the people here?	95 yes	90	100	91	100	*
Are the staff helpful?	96 yes	90	100	100	83	*

* Nonsignificant difference between groups; p is significant only when less than .05.
** Significance levels were determined from chi square analysis of the four continuum groups. Value p denotes the odds of a particular result being found by chance.

not prove to be a good predictor of where each home would fall along the contiuum. One home at the normalized end of the continuum was a nuclear family with six school-age children; this instrument was not appropriate for their level of development and maturity. Because of this fact the findings cannot be clearly interpreted.

Staff Interviews

One staff member at each group home who had direct, daily contact with residents was interviewed. This person's answers to the group home questionnaire determined general demographic information about the facility and the program. A second instrument, the Resident Management Practices Inventory, focused on four primary indices: rigidity, block treatment, depersonalization, and social distance.[23] The more rigid the routine of the home, the more management practices are considered institution-

al. Conversely, such practices are more normative and resident-oriented when they are flexible according to individual differences among residents or circumstances.

Block treatment means that the residents are treated as a group rather than individuals before, during, or after any specific activity. Depersonalization in group-home practices may be seen when there are no opportunities for residents to have personal possessions or privacy, and no outlets for self-expression or self-initiative. When there is a sharp separation socially between staff and residents, the management practices are considered to be more institutionally oriented. Residents are more normalized when the social distance is reduced, by staff and residents sharing living spaces and interacting in functionally diffuse and informal situations.

The short form of the Community Oriented Programs Environment Scale (COPES) was administered to each primary staff member interviewed.[24] The focus of the scale, which was developed by Rudolph Moos at Stanford Univer-

23. R. D. King, N. Raynes, and J. Tizard, *Patterns of Residential Care: Sociological Studies in Institutions for Handicapped Children* (London: Routledge and Kegan Paul, 1971) p. 78.

24. R. H. Moos, *Community-oriented Programs Environment Scale* (Palo Alto, CA: Veterans Administration, March 1973).

TABLE 1-5. RESIDENT MANAGEMENT PRACTICES INVENTORY: STATISTICAL SUMMARY

Index	Total Group	Breakdown of the Four Continuum Groups			
		Institutional Group	Low Medium Group	High Medium Group	Normally Supportive Group
Rigidity	3.94	4.66	3.66	3.75	4.00
Block Treatment	2.00	2.33	2.16	1.00	1.75
Depersonalization	5.88	5.33	5.66	6.00	6.50
Social Isolation	5.19	5.66	5.16	6.00	4.50
Total Score	16.94	18.00	16.66	16.75	17.00

Data reported are means for each group. All findings resulted in nonsignificant differences between the institutional and normally supportive group as determined by analysis of variance.

Some sample questions from this instrument are:
• Do residents have to get up at the same time every day?
• Can residents go into their bedrooms anytime they wish and stay there if they wish?
• Are there certain times for bathing?
• Are residents allowed to fix up or decorate around their beds or in their rooms?
• Do residents prepare meals?
• Do staff on duty eat with residents?

sity, is the interaction of people and their environment.

The COPES instrument includes statistical norms derived from administering this instrument in a variety of community programs across the country. The results of our study fall within those statistical norms, indicating that the homes selected were similar to those in the broader community.

The COPES is conceptualized along ten primary subscales.

1. *Involvement* measures the activity of members in the day-to-day functioning of their program.
2. *Support* measures the extent to which residents are encouraged to be helpful and supportive and the extent of staff support.
3. *Spontaneity* explores the extent to which the program encourages members to act openly and express their feelings freely.
4. *Autonomy* assesses the residents' levels of self-sufficiency and independence in decision-making and relationships with staff.
5. *Practical Orientation* measures the extent to which the environment orients a member toward preparing for release from the program.
6. *Personal Problem Orientation* measures the extent to which members are encouraged to be concerned with their personal problems and feelings and seek to understand them.
7. *Anger/Aggression* measures the extent to which members are allowed and encouraged to argue with members and staff.
8. *Order and Organization* measures these qualities in terms of residents' personal appearance, house maintenance, and staff encouragement of order.
9. *Program Clarity* measures the extent to which the member knows exactly what to expect in the day-to-day routine of the program.
10. *Staff Control* assesses the extent to which the staff uses rules to keep members under necessary controls, and how well members and staff relate.

Daily Activity Questionnaire

We developed this questionnaire specifically for our study to investigate how various spaces in group homes were being used, what problems were present, and what types of approaches were used by each group to resolve these problems. Areas of inquiry spanned five broad topics defined by space and function: Kitchen and Meals, Security and Safety, Bathroom, Bedroom/Personal Space, and Free Time Spaces (living room, game room, dining room).

The questionnaire was given to each resident as well as to the primary staff member interviewed at each group home. Results of the first four parts of the questionnaire are given in table 1-7.

In the free-time spaces section of the questionnaire, residents were asked to respond to the questions by picking from the following list of rooms: large living room, small living room, game room, dining room, kitchen, bedroom, hallway, office, outside of the house, other. Responses in table 1-8 are ranked from A to C, with A representing the most frequent response. Responses are from the total survey.

TABLE 1-6. COPES: STATISTICAL SUMMARY

Index	Study Data		Normative Data	
	Mean	s.d.	Mean	s.d.
Involvement	3.29	.92	2.38	.83
Support	3.47	.94	3.29	.54
Spontaneity	2.94	.75	2.51	.70
Autonomy	2.27	.59	2.60	.71
Orientation	2.88	.72	2.99	.56
Personalization	3.06	.97	2.71	.86
Anger	2.80	.94	2.57	1.02
Order	1.93	1.07	2.27	.89
Clarity	3.35	1.06	3.19	.60
Control	2.19	.98	1.63	.78
Relationship	9.71	2.09	8.38	—
Treatment	10.29	2.82	10.87	—
Administration	7.65	2.78	7.09	—
Total COPES	28.36	2.82	26.34	—

TABLE 1-7. DAILY ACTIVITY QUESTIONNAIRE: STATISTICAL SUMMARY

KITCHENS AND MEALS	Response	Total Survey % Response	Institutional Group % Response	Low Medium Group % Response	High Medium Group % Response	Normally Supportive Group % Response	p**
			Breakdown of the Four Continuum Groups				
Do people in your house eat together as a group?	Yes	98	97	95	96	98	*
Which meals?	Breakfast and dinner	68	59	72	71	67	*
Do you enjoy eating as a group?	Yes	90	95	86	90	92	*
Do you do any grocery shopping for the house?	Yes	52	78	19	57	57	p < .008
Who makes out the menu and shopping list?	The staff	77	86	90	88	44	p < .04
Do you do any cooking for the house?	Yes	59	37	52	60	91	p < .004
Do you know how to work the appliances?	Yes	89	74	97	81	100	p < .01
Do you eat all your meals in one particular room of the house? The dining room?	Yes	69	100	65	68	52	p < .01
Are there other rooms besides the dining room where people eat? The kitchen?	Yes	54	63	55	58	76	*
Who cleans up after dinner? Rotated chore among residents?	Yes	51	41	36	65	67	p < .04
Who does the dishes? Rotated chore?	Yes	60	61	52	62	71	*
Is there room enough in the kitchen to talk with or watch others cooking?	Yes	81	83	77	77	91	*
SECURITY AND SAFETY							
Do you have your own key to the house?	No	65	68	68	73	45	*
Does someone make sure that the door is locked at night? The staff?	Yes	86	100	87	88	68	p < .04
Do you feel safe and secure in this house?	Yes	94	95	90	96	100	*
Are there any safety problems in the neighborhood?	No	61	59	71	64	45	*
While you have been a resident here, has this house been broken into?	No	92	100	84	92	95	*
BATHROOM							
Do you generally use the same bathroom all the time for washing, bathing, etc.?	Yes	88	74	90	100	81	p < .04
Do you have to wait because someone else is using the bathroom?	No	57	68	61	42	62	*
Who cleans up the bathroom?	Rotated chore	47	57	57	43	58	*
	Each user	26	14	30	43	26	*
Where do you keep your soap, towels, toothbrush, etc? Own room?	Yes	60	79	52	73	38	*
Do several people use the bathroom at the same time or is it always only one person at a time?	Yes	67	67	90	62	40	p < .002

TABLE 1-7—CONTINUED.

BEDROOM—PERSONAL SPACE	Response	Total Survey % Response	Breakdown of the Four Continuum Groups				p**
			Institu- tional Group % Response	Low Medium Group % Response	High Medium Group % Response	Normally Supportive Group % Response	
Where do you go when you want to be alone?	Bedroom	92	72	95	100	100	p < .04
Do you share a bedroom?	Yes	45	5	81	56	19	p <.001
	No	54	95	19	44	81	p <.001
Do you prefer having your own room?	Yes	76	95	60	64	95	p <.004
Who does the furniture in your bedroom belong to? The house?	Yes	76	60	97	88	42	p <.001
Do you like your room?	Yes	95	94	93	96	95	*
What have you used to decorate your room?	Pictures	24	19	16	21	28	*
	Nothing	43	62	41	56	40	*
Do you have enough space for storing your clothes and other possessions?	Yes	88	84	81	96	91	*
Do you do your own laundry?	Yes	65	26	84	73	62	p <.002
Do people knock before entering your bedroom?	Yes	72	89	68	62	76	*
Is there a lock for your bedroom door?	No	60	35	74	50	70	p < .03
If there is a lock, do you have a key?	No	82	73	100	58	94	p <.002

* Nonsignificant difference between groups; p is significant only when less than .05.
** Significance levels were determined from chi square analysis of the four continuum groups. Value p denotes the odds of a particular result being found by chance.

TABLE 1-8. FREE-TIME SPACES QUESTIONNAIRE

1. Where do you spend most of your free time?

    A. Large living room
    B. Bedroom
    C. Outside

2. Where do you spend most of your day?

    A. Large living room
    B. Outside
    C. Other

3. Which room is the center of activity in the house?

    A. Large living room
    B. Dining room
    C. Kitchen

4. Where are group meetings held?

    A. Large living room
    B. Dining room
    C. Office

5. Which is the noisiest room?

    A. Large living room
    B. Kitchen
    C. Dining room

6. In which room do you watch television the most?

    A. Large living room
    B. Bedroom

7. Where do you listen to records or the radio?

    A. Bedroom
    B. Large living room
    C. Game room

8. Where do you visit with friends or relatives?

    A. Large living room
    B. Outside
    C. Small living room

TABLE 1-8—CONTINUED.

9. Which room is used the least?

   A. Other
   B. Small living room
   C. Game room

10. Which is your favorite room?

   A. Large living room or bedroom (tie)

11. Where are books, magazines, and newspapers kept?

   A. Large living room
   B. Small living room
   C. Dining room

12. Are there any rules or regulations for using these rooms?

   A. Yes (for all choices)

13. Who cleans these rooms?

   A. Rotated chore (for all rooms except bedroom)

14. Who arranged the furniture in these rooms?

   A. Staff (for all rooms)

15. Do you ever want to rearrange the furniture in these rooms?

   A. No (for all rooms)

16. Where do you work on your hobbies?

   A. Bedroom
   B. Office
   C. Other

17. Where do you keep your plants?

   A. Bedroom

Direct Observation

In each group home, direct observations of the use of space, activities, and interactions were conducted. The method used for these observations was behavioral mapping, a technique developed by environmental psychologists for relating various aspects of behavior to the physical spaces in which they occur.[25] One "map" was completed every five minutes during the half hour before the evening meal, during the meal, and the half hour following. Each staff member and resident who was observed on the first floor of the home during the assessment period was "mapped." The information recorded included which room they were in, the type of activity they were engaged in (talking, reading the paper, watching television, eating), the level of social interaction, and verbal expression.

Such observations revealed that residents typically tended to take on a passive, consumer role. Before dinner, the staff are actively involved in meal preparation in the kitchen, while residents are waiting for dinner in the living room—most often watching television and not interacting with others. During meals, the activity moves to the dining room, with the major focus on eating—not social interaction. Most residents seem to follow an "eat-and-run" approach while staff try to engage the residents in conversation. After dinner the staff tend to remain in the kitchen and dining room, while residents move into the living room for more noninteractive television viewing.

Some interesting differences were also noted along the institutional-to-normalized continuum through behavioral mapping. In homes at the normalized end of the continuum, residents were found to participate in more social interaction with others, talking much more among themselves and engaging in cooperative activities, and there was a much lower incidence of passive television watching than in homes toward the institutional end of the continuum. As opposed to the average group-living experience, those in normal environments tend to stay in the dining room

25. William Ittelson, Harold Proshansky, and Leanne Rivlin, *Environmental Psychology: Man and His Physical Setting* (New York: Holt, Rinehart and Winston, 1970), pp. 658–68.

1-1. Mapping form used to record observed behavior.

longer, perhaps lingering over coffee, talking and enjoying each others' company.

Direct observation also included a building and neighborhood assessment. This inventory, designed by the research team, was intended to survey particular features of the physical facility. The information gathered included age of the building, number of rooms, their arrangement and use, accessibility for wheelchairs, changes made to comply with codes, and location within the neighborhood. Floor plans of the entire house were sketched, including furniture arrangements. Interior photographs were taken of typical bedrooms (after first obtaining permission from the resident), all first-floor rooms, and any special or unique features.

Mail Survey

In order to investigate the premise that the survey homes were representative of group living generally, extensive questionnaires were mailed to approximately two hundred group-living facilities in Ohio. These facilities included homes administered by the Veterans Administration, sponsor care homes that receive a daily rate for each resident, and agency-supported community facilities funded by county and state sources. A 32 percent return rate of questionnaires was achieved.

The mailed survey contained many of the same questionnaires and scales used in the survey homes. The primary purpose of these "mailers"

TABLE 1-9. BEHAVIORAL MAPPING: STATISTICAL SUMMARY

Variable	Total Survey Mean %	Breakdown of the Four Continuum Groups				p**
		Institutional Group	Low Medium Group	High Medium Group	Normally Supportive Group	
PERSON						
1. Resident	76	78	79	71	76	*
2. Staff	17	12	16	22	21	*
LOCATION						
1. Large living room	30	39	28	25	26	*
2. Small living room	7	2	14	6	3	*
3. Dining room	25	12	23	30	43	p < .04
4. Kitchen	22	18	22	22	25	*
5. Hall	9	13	11	8	2	*
6. Recreation room	3	8	0	3	0	*
7. Other room	5	8	3	6	2	*
ACTIVITY						
1. Cooperative activity	25	17	25	27	37	p < .05
2. Solitary activity	9	3	15	10	9	*
3. Moving, walking, in transit	9	16	6	8	2	*
4. Meal-related activity	14	13	10	18	13	*
5. Eating	17	14	15	20	18	*
6. Watching TV	18	40	24	20	16	p < .04
7. Talking on the telephone	2	1	2	3	2	*
8. Passively watching others	7	10	12	2	4	*
LEVEL OF INTERACTION						
1. Cooperative activity, within 3 feet of other person(s)	27	18	17	42	36	p < .03
2. Cooperative activity, more than 3 feet from other person(s)	15	6	22	16	16	p < .05
3. Parallel, within 3 feet of other(s), but not engaged in cooperative activity	35	56	36	16	28	p < .01
4. Solitary, more than 3 feet from other(s), not engaged in cooperative activity	23	20	25	26	20	*
LEVEL OF INTERACTIVE SPEECH						
1. Talking with other(s)	41	25	36	53	56	p < .03
2. Talking alone, using telephone	3	1	6	3	1	*
3. Hostile, yelling, arguing	1	0	2	1	0	*
4. No speech	55	74	57	42	43	p < .02
FOCUS OF INTERACTION						
1. Interaction with another resident	32	16	45	33	37	p < .03
2. Interaction with staff	11	7	8	15	16	*
3. Interaction with other, visitor	6	4	5	11	7	*
4. No interaction	51	74	42	42	40	p < .02

* Nonsignificant difference between groups; p is significant only when less than .05.
** Significance levels were determined from chi square analysis of the four continuum groups. Value p deontes the odds of a particular result being found by chance.

TABLE 1-10 BUILDING AND NEIGHBORHOOD ASSESSMENT

1. Building age: range 11 to 100; average 64
   Previous use: (13)Home
               (0)Apartment
               (0)Commercial
               (2)Other: dormitory, convent

2. Changes made to comply with codes:
   (3)Enclosed stairways
   (0)Installed sprinklers
   (8)Added fire extinguishers
   (6)Electrical changes
   (8)Added smoke detector/alarm
   (6)Added fire escapes
   (3)Installed exit signs
   (1)Installed partition walls
   (2)Added bathrooms
   (1)Other: new furnace

3. Is your place accessible to wheelchairs?
   (0)Yes   (15)No

4. Are the outside spaces of the house used?
   (12)Recreationally   (5)Gardening
   (8)Casually   (2)Not used

5. Which community setting do you consider your house to be a part of?
   (7)Residential neighborhood
   (5)Edge of neighborhood
   (0)Business community
   (1)Rural
   (0)Isolated
   (2)Other: university or institution

6. How close is your home to grocery shopping, drug stores, and variety stores?
   (9)Within walking distance   (1)Twenty-minute ride
   (4)Ten-minute ride   (1)Half-hour ride

7. How close are the nearest places for recreation and entertainment?
   (7)Within walking distance   (3)Twenty-minute ride
   (4)Ten-minute ride   (1)Half-hour ride

8. How close are the nearest banking and postal services?
   (6)Within walking distance   (2)Twenty-minute ride
   (6)Ten-minute ride   (1)Half-hour ride

9. Are you near a public transportation line?
   (11)Yes   (4)No

was to identify common populations served and to compare the group living experience statewide to the preselected fifteen group-living settings in the detailed study. Analysis of the results

TABLE 1-11. DIFFERENCES BETWEEN TYPES OF GROUP HOMES

	Site Visited	Surveyed by Mail		
	Homes	Agency Homes	VA Homes	Sponsor Homes
Number	15	6	33	32
Mean age of residents	31	36	43	42
Mean length of stay	1–6 months	1–6 months	3–5 yrs.	1–3 yrs.

revealed that the Veterans Administration homes and sponsor care homes served an older, more chronic, long-term resident; this was a different population from that served by agency-supported homes and from the detailed study. From this data it can be seen that the homes visited and those agency homes from the mailed survey serve a younger population in a more short-term rehabilitation program. The findings of this study may most directly reflect community group-living settings that emphasize growth and rehabilitation for clients through short-term programs.

Common Problems

To picture the common situations and problems of group living in the homes surveyed, we created a composite resident, group home, and program. This imaginary picture is, in one sense, an average of numbers and should be viewed with some caution. The composite resident is a man, while it is obvious that there are women in group homes; the average age is thirty-one years, but the range in group homes surveyed was from twelve to seventy. The picture of the composite home and the composite program are both also widely varied, and none is really "typical."

In addition to being a thirty-one-year-old man, the composite resident is single and lives with eight other men and women. He has moved from a psychiatric hospital during the past six months, and his goal is to gain sufficient personal resources to be independent and live alone in an apartment within the larger community.

He is generally satisfied with his current situation and gets along well with other residents. He feels comfortable, safe, and well fed and finds the staff in the house helpful. He maintains his own room and is responsible for some personal as well as shared chores. He does not have any control over the budget for the house and has few personal assets. In a society that places high value on productive competitive employment and wage earning, his income is less than $3000 per year. Even though he has had prior work experience and training in a skill, he now works at an unskilled job.

In his spare time, he listens to the radio or watches television, rarely becoming involved in outside community activities or social events. His closest friend does not live at the group home, but because he does not entertain guests at the group home, he sees his friend infrequently.

The composite group home accommodates nine residents. The house itself is about sixty years old and was previously a private home. A few renovations were made to comply with fire and safety regulations but not to make the house accessible to people in wheelchairs. The house has two floors with all of the standard spaces but more bedrooms than might be considered typical. It is adequately but not luxuriously furnished, reasonably well cleaned and maintained. Not surprisingly, it shows signs of wear and heavy use.

The home is located near the edge of a residential neighborhood, not in the middle of a block. Shopping, recreation, banking, and other services are within walking distance, and public transportation is accessible.

The composite program has four full-time, two part-time, and three consultant staff members. Eighty-two percent of the residents stay in the house for a period of less than twelve months; slightly less than half stay less than six months.

Because programs are so varied, it is difficult to provide one composite example. A major difference from place to place is the amount of time residents spend in the house and away from it. In-house programs concentrate on living skills, social skills, and personal problems. Outside programs are either vocational or educational in nature. Some houses simply provide care and boarding with no structured program.

Residents are responsible for getting themselves up and ready for the day. Two meals are eaten as a group—either breakfast and supper or lunch and supper. Residents are also expected to do routine house-cleaning chores. Most program activities occur during the day, and evenings are generally considered free time. Some programs encourage residents to be involved in independent activities, while others provide a highly structured schedule.

From the survey and the imaginary picture of the composite resident, home, and program, three common problems emerge: Residents lack a sense of control and ownership in the group home; privacy, whether for one person or a social group, is difficult to attain; and the shared spaces are too often limited to one activity and are not supportive of a wide range of social experience.

The issue of ownership and control is sometimes difficult to grasp. It relates to residents' self-image and self-esteem, and with whether residents feel part of the fabric of the house or just transient factors. They do not see themselves taking an active part in setting programmatic rules and goals, nor do they have a sense of personal ownership or territory in the home. This division of roles permeates the group-living experience and is most obvious in the division of labor. Residents are responsible for such chores as setting the table, washing dishes, dusting, and other light cleaning chores, but they do not plan menus, shop for groceries, repair leaky faucets, or make decisions about repainting rooms. These duties are under the control of staff members.

Although the residents make heavy use of the living room, they are not involved in making the regulations for its use—rules about smoking, scheduling hours of use, or visiting. The staff makes the rules. Even in such matters as furniture arrangement, residents rarely have a say. Surprisingly, even in the bedrooms, most residents do little to control the space. Although they are permitted to do so, they do not decorate and personalize, and there seems to be little encouragement in this direction.

Privacy is a highly desirable but elusive state in most group homes. Residents report that their bedroom is the place of choice to be alone. Those who share bedrooms are not so assured of finding privacy and, on the whole, would prefer a single room.

Privacy for visiting or being with just one other person also presents a difficulty. Bedrooms are generally considered off-limits for such purposes, and there are few other small places in most homes that can be made private.

The limited uses of shared space, the third common complaint, can be solved with imagination. In most houses shared spaces are set up for only one activity. Potentials for having separate activities occurring in the same room are missed.

One such missed opportunity is often in the dining room arrangement. Although mealtime is the only time the whole group uses it, it is often permanently arranged to accommodate everyone at the group meal. The rest of the day, it is unused and empty. The living room too is set up usually for television viewing, though rarely does the group watch as a group. Kitchens and dining rooms are used only for meal-related activities.

The social experiences that are consciously supported are those where the entire group comes together: house meetings and group meals. Little support is given for smaller groups of two, three, or four people. Furniture arrangements, which can define how space is used, are not set up to support casual interactions.

When looking at the houses as a whole, the overall picture is that residents see themselves as living in someone else's home, with little sense of belonging or feeling of pride. Passivity and dependence are unconsciously supported. The range of social experiences is limited.

Again, remember that the three common problem areas are drawn from an imaginary picture of a composite resident, place, and program. These problems are evident to some degree in all the homes surveyed. In places where normalization is stressed, there is more productive activity, positive social interaction, and a minimum of passive, isolated behavior. In general, such homes deal more successfully with these problems than do other homes.

Sites Surveyed

Building plans are useful aids for seeing each house. Examine them with these elements in mind: privacy, variety, domestic scale, and effective circulation. The building plan is only half the story, however. Furniture can be arranged to support many activities in one space, to put people at comfortable conversation distances, and to make a house inviting.

The fifteen group homes surveyed are each represented here by a description and a ground floor plan. Each plan is drawn to the same scale for comparison purposes.

Figure 1-2 shows the floor plan of a house operated as a "lodge" for fifteen adult men and women who have experienced long-term hospitalization or repeated admission. This is one of two houses that stand side by side; the second house is used primarily for bedrooms, ensuring all lodge members have their own private bedrooms. Meals are served in the main house only, although there is a snack kitchen in the second house. The houses are located on a main street at the edge of a residential neighborhood in a large urban area. In addition to running these houses, the lodge operates (at a separate location) two businesses that employ the residents. The program emphasizes providing a place to live and an income for residents immediately upon discharge from the hospital. One staff person works in the house eight hours a day, preparing meals and generally overseeing the operations of the house. Residents are responsible for some day-to-day chores and for their own medications and appointments at the local mental health center. (After the initial survey, this home was remodeled with some interesting changes in the use of space as a result. These changes are discussed in sections 4 and 5; this plan shows the house as it was at the time of the survey.)

The plans for a halfway house are shown in figure 1-3. The house has ten residents—adult women and men who have had long-term mental hospitalization or repeated admissions. It is an old house in a mixed-use zone of a large urban area. This program emphasizes daily living and survival skills. Two staff members are in the

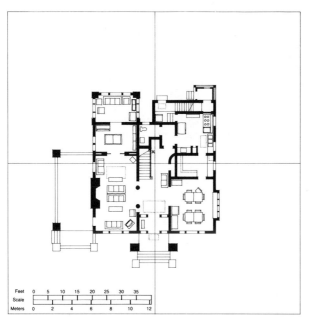

1-2. Floor plan for 15-resident lodge for mental-health patients.

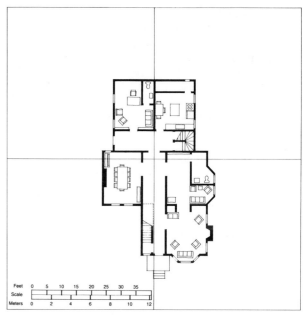

1-3. Floor plan for 10-resident halfway house for mental-health patients.

1-4. Floor plan for 5-resident house for a commune.

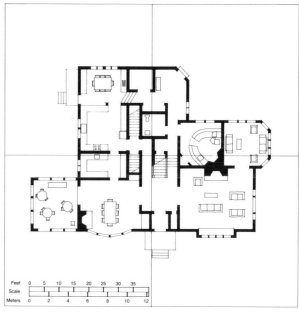

1-5. Floor plan for 8-resident house for a nuclear family.

home during the day shift; the residents manage affairs themselves the rest of the time. The staff is involved primarily in teaching living skills and in seeing to administrative matters. Residents are responsible for all meals, cleaning, and other day-to-day affairs.

Figure 1-4 shows a cooperative household for five adult women and men. It is located in an average-sized older home in a residential suburb of a large urban area. This communal group (a nonidentified group with no connection to the mental health field) was included in the survey to give a balanced picture of the problems of group living. Shared habits, attitudes, and commitments to a counterculture lifestyle are the common denominator of the group. Members all follow a vegetarian diet, play musical instruments, watch no television, and smoke no tobacco.

Figure 1-5 is a plan of the home for a large nuclear family of eight members: parents and teenage children. This typical, though upper-income, family was included in the study to help understand the problems that occur when eight people share space. The family resides in a very large older home in an upper-income suburb. Family spirit and commitment to mutual support are pronounced. The children all share in meal preparation, laundry, cleaning, and other house-hold chores, as would be typical for most families.

The plans for a cooperatively owned large house for five adult women, located in an older suburb, are shown in figure 1-6. This group of women has a strong commitment to living coop-eratively and sharing the responsibilities of own-ing and running their house.

Figure 1-7 shows a halfway house for twelve adult men prisoners serving the remainder of their sentence in a community setting. The house is located near the edge of an older neighborhood next to the downtown of a medium-sized urban area. Staff people are on duty around the clock to monitor activities and provide support. The program emphasis is on the residents making their own individual plans and carrying them through, encouraging the residents to regain con-trol over their own lives.

In figure 1-8, a home for five teenage girls with behavioral problems is shown. It is located in a small ranch-style house in a somewhat isolated semirural area. The live-in houseparents receive

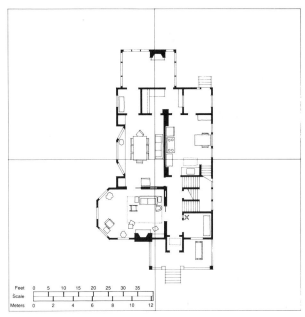

1-6. Floor plan for 5-resident house for a co-op.

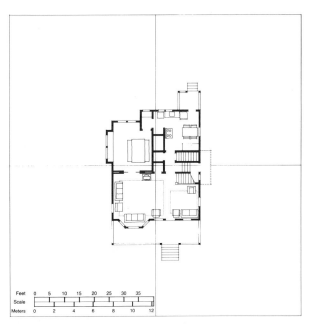

1-7. Floor plan for 12-resident halfway house for prisoners.

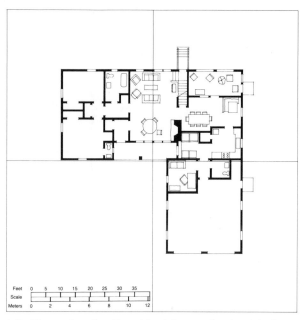

1-8. Floor plan for 5-resident house for troubled juveniles.

professional guidance from a county mental health agency. This tightly run program is oriented toward behavior modification by providing a stable and structured environment. The girls attend regular schools during the day.

Figure 1-9 shows a former dormitory on a state university campus, which houses a program for fifteen adult men and women. Operated by a nearby long-term state mental hospital, the program is based on the Fairweather model of a self-governing group. In the Fairweather model, residents make all decisions and are responsible for the day-to-day operation of the building. Residents establish their own system of rewards and penalties for completion or neglect of assigned chores and responsibilities. Staff people are on duty full time, but their interaction with residents is kept to a minimum.[26]

26. George W. Fairweather, editor, "The Fairweather Lodge: A Twenty-five Year Retrospective," *New Directions for Mental Health Services*, no. 7, 1980, pp. 13–33.

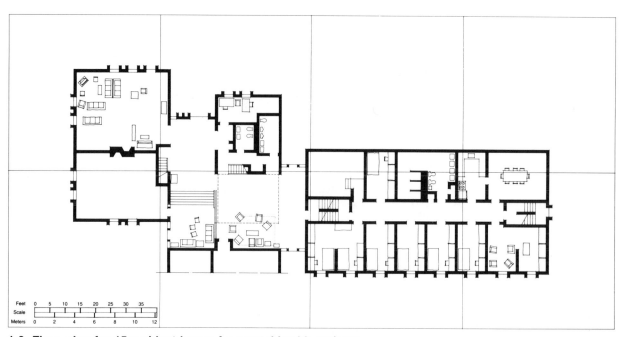

1-9. Floor plan for 15-resident house for mental-health patients.

The floor plan in figure 1-10 shows an old farmhouse that is now a home for six retarded adult men. During the day, the residents go to a sheltered workshop. The houseparent couple that runs the house works individually with residents on personal goals and living skills and concentrates on helping them learn to handle everyday situations.

Figure 1-11 shows a home for eleven adult men with alcohol- or drug-related problems. It is located in a residential neighborhood near the downtown of a small urban area. Residents are generally court-referred and stay an average of ninety days. The program provides full-time staff and a controlled living environment in conjunction with local Alcoholics Anonymous programs. Several residents continue to hold their regular jobs.

1-11. Floor plan for 11-resident house for alcohol and drug abusers.

1-10. Floor plan for 6-resident house for the retarded.

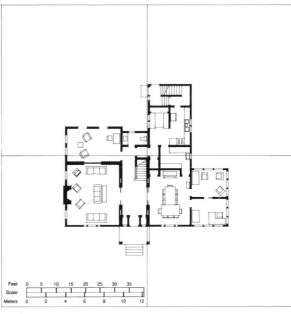

1-12. Floor plan for 12-resident house for mental-health patients.

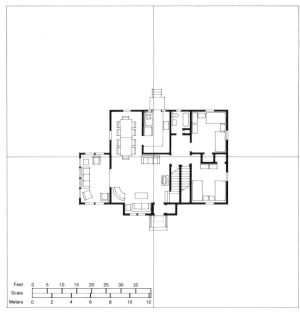

1-13. Floor plan for 10-resident house for mental-health patients.

1-14. Floor plan for 9-resident house for mental-health patients.

Figure 1-12 is a floor plan for a halfway house for twelve adult men and women who have been or would otherwise be in psychiatric institutions. The house is a large old mansion in a mixed-use area, somewhat isolated from residential neighborhoods. The program and full-time staff emphasize individual vocational or educational planning and strive to place people in a school or job in an area of interest.

Figure 1-13 shows one of two homes that are part of a predischarge program for residents of a psychiatric institution. Ten adult men and women live in the house, which is located on the grounds of a long-term psychiatric facility and was formerly used as a doctors' residence. The program emphasizes individual planning—with support of a full-time staff—for the resident's own discharge process and future living arrangements.

The plan in figure 1-14 is for a sponsor care home for nine older women and men, former residents of psychiatric institutions. In a sponsor care program, the operator receives a per diem reimbursement from the state for each resident. The small building shown, once a convent, is in a residential section of a large urban area. The residents participate in a day program at a nearby mental health center. A housekeeper provides three full meals a day and runs the house as a boarding and care program.

Figure 1-15 shows an old mansion located at the edge of a residential neighborhood in a large urban area. It is part of a mental health transitional program, housing twenty-two generally young adult men and women who were referred from a nearby short-term mental hospital. Residents are divided into two "family" support groups to deal with their individual and group problems and to plan for their future. There is full-time staffing. A major thrust is the prevention of repeated admissions.

Figure 1-16 is a plan for a sponsor care home for five adult men, former residents of psychiatric institutions. This large old house on the fringe of a residential neighborhood is owned by the operator, who does the cooking, cleaning, and other chores and teaches some living skills. Residents occupy the second floor only. This is primarily a boarding and care program.

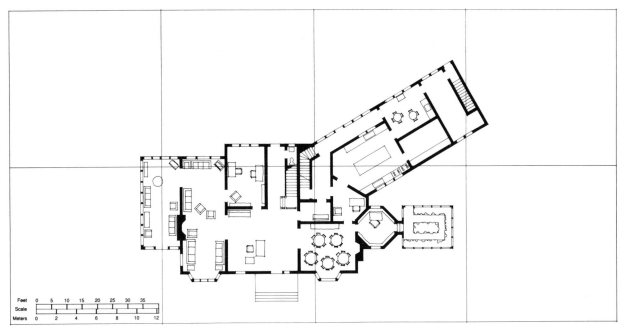

1-15. Floor plan for 22-resident house for mental-health patients.

Floor Plans and Character

The subtle differences between the geometry of houses make a noticeable difference to the overall use of space and the success of a group-home program. In judging a house, three concerns must be addressed—layout and its effect, circulation patterns and their effect on a living environment, and the unique physical features in a house that facilitate living together as a group. The number of bedrooms and the kind of carpeting will not determine the suitability of a house as a group home.

The houses we surveyed can be divided into three types according to floor plan. In the first type, shared spaces on the first floor are arranged around a central hallway (fig. 1-17). These houses are generally located on a spacious

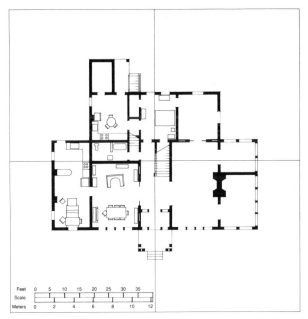

1-16. Floor plan for 5-resident house for mental-health patients.

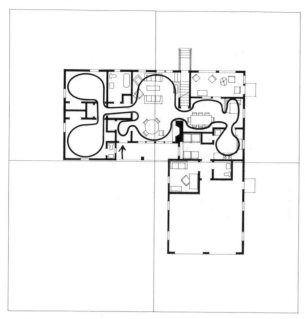

1-17. Floor plan of an older house showing shared spaces arranged around a central hallway.

lot and have a more formal, symmetrical character than the others. The second group has rooms strung in a row in an elongated building that fits into a narrow city lot (fig. 1-18). The third set, built recently, has a "modern" (open plan) layout, no hallway circulation, a small entryway, and rooms that open one to another (fig. 1-19); they are less formal than the first two building types.

There are advantages and disadvantages to each type. The third group of postwar houses with open layouts has fewer of the equipment and maintenance problems than the older houses, and they typically operate on much lower energy cost. However, some of them, depending on climate and because of their low ceilings and lack of cross ventilation, can be comfortable in hot weather only if air conditioned, which is costly and conducive to keeping residents indoors. Many have circulation patterns through the middle of rooms, abrupt entry into living spaces, and little privacy. Some of the useful features of older houses are missing: real front porches, adequate storage space, sliding doors, high ceilings.

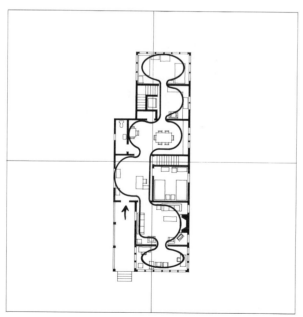

1-18. Floor plan of an elongated house with rooms strung in a row.

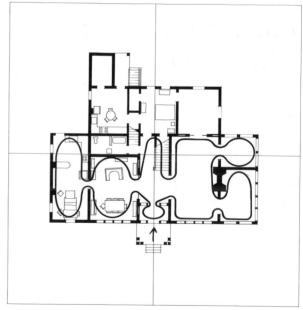

1-19. Floor plan of a contemporary house with less formal open-plan layout.

The narrow, elongated houses in group two often provide numerous private bedrooms that are advantageously located away from social areas. Unfortunately, circulation is often through the central shared spaces, making it difficult to furnish or use them comfortably.

Houses in group one often have large rooms that are difficult to work with. A large living room does not function well as a single social space, because furniture arrangement—often positioned against the walls—leaves people too far apart for good interaction. Yet with imagination and a little effort smaller social areas can be created in a large room. Very large bedrooms are usually shared by too many people or, to avoid this, are sometimes partitioned. Either way, the original quality and scale of the rooms are destroyed. On the positive side, the central hallway allows circulation through the center of the house, leaving shared spaces on the first floor unbothered. Rooms can be closed off occasionally for privacy without stopping all movement through the house.

In houses with similar floor plans, small layout differences can make one house more livable than another. The houses in figures 1-2 and 1-12 both have large, central entry halls with living and dining rooms on either side and similar kitchen and pantry locations. At a glance the plans look almost identical. However, there are differences in circulation and separation of areas, occurring because of the way the houses were built or are used (see figs. 1-20 and 1-21).

In the house in figures 1-2 and 1-20, the sun room, the most private shared space, can only be reached by walking through the heavily used living room and then through a small music room. Because of this traffic, the furniture arrangement of the living room is limited, and activities there are often interrupted. It also makes it impossible for someone to reach the sun room without being seen, an often discouraging fact. The impression created by the social spaces, as a whole, is a string of interconnected spaces, all within view of the living room where most people congregate.

In contrast to the living room, the kitchen, when the survey was made, had only one entryway from shared spaces and no place for people to sit and talk. Ironically, this made it a sanctu-

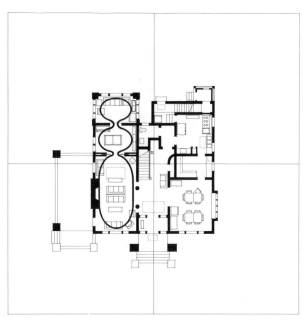

1-20. Multientry rooms, as highlighted in this floor plan, make obtaining privacy difficult.

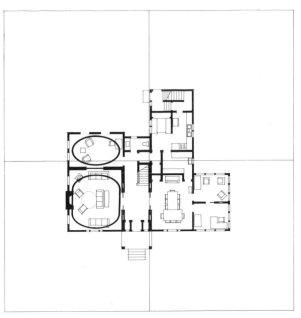

1-21. Single-entry rooms, as highlighted in this floor plan, allow for more private "pockets" of activity.

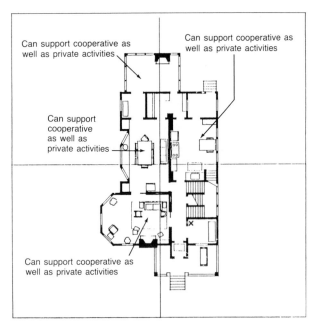
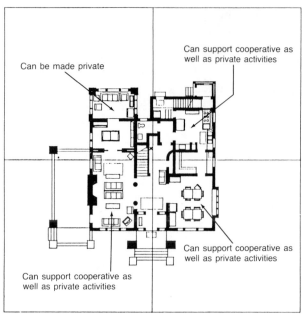

Can support cooperative as well as private activities

Can support cooperative as well as private activities

Can support cooperative as well as private activities

Can support cooperative as well as private activities

Can be made private

Can support cooperative as well as private activities

Can support cooperative as well as private activities

Can support cooperative as well as private activities

1-22. Shared spaces should encourage both individual and group activities.

ary for the cook rather than a place for residents to gather, even though the staff wanted it to be used as a casual gathering place. Since the survey, the room has been remodeled and has become the inviting space the staff envisioned.

In the house in figures 1-12 and 1-21, the kitchen works as a "backstage" place where formalities are relaxed.[27] It is easily reached by one of two passages, and a comfortable booth is especially inviting.

Because there is only one entrance to the living room and the library in the house shown in figures 1-12 and 1-21, there is no traffic problem. People do not interrupt each other as often, and furniture can be arranged to contain small pockets of activity—in kitchen, office, living room, and library—that feel separate. Rather than the whole group tending to congregate in one room, this floor plan allows for several parallel activities—group or individual—without one disrupting the other.

One of the advantages of selecting an older

home rather than building a new one is that it is already integrated into a neighborhood, lessening the likelihood of stigma. In newly built houses, residents are often perceived as different simply because their house is new to the neighborhood and so are they. Because the demand for large, older homes is on the wane today, it is usually easy to find one that adapts well to group living. The cost of building a comparable new house generally exceeds the cost of an existing one. A requirement for state funding in Ohio, which is typical of several states, is that purchase and renovation costs of an existing house not exceed 70 percent of costs for an equivalent new structure. Whenever costs begin to exceed this figure, the economies would favor new construction. The extent of the remodeling required may decide this. Remodeling costs, in turn, depend on the program, future residents, and fire and building codes.

A very important recommendation for an older house is its richness of detail and atmosphere. New buildings, including those designed to be group homes, are usually commonplace, with box-shaped rooms and no particular features that suggest active use. Older homes have geometries

27. Erving Goffman, *The Presentation of Self in Everyday Life* (Garden City, NY: Doubleday Anchor Books, 1959), p. 112.

that do not often occur in new construction, from high ceilings and wide staircases to bay windows, window seats and good cross ventilation. Many include arched openings, L-shaped and T-shaped rooms, nooks and alcoves, and such details as intricate wood moldings, cornices, fireplaces, mantels, and hardwood floors. They have a sense of place and history and are, therefore, less likely to be cold, sterile, impersonal, and untouchable than new homes.

Shared Spaces

How can a group of peers with different backgrounds, needs, and interests share a space while maintaining an individual sense of ownership? How can people maintain their right to choose solitude or socializing? Residents in many group homes are relatively transient, and the process of developing communalities and satisfying individual needs is a constant one. Even in permanent groups it requires ongoing attention.

Shared spaces can belong to residents, members of the staff, the home operator, to all of these, or to none. The most visible shared spaces in a group home are the living room, den, and dining room. Less obvious but, with a bit of imagination, just as usable are hallways, corridors, and stairways; kitchen, laundry room, and even bathroom. Shared space can offer chances to be alone, to converse or meet with a large group. Several activities can occur at the same time in one house if all available spaces are used advantageously.

Survey houses at the more normal end of the continuum have a large number of shared spaces to choose from. Comparison among survey houses was based on the number of spaces that could be used for group activities, for private individual activity, and for two different activities at the same time. The four houses that our measures rated normal have significantly more usable space than the three houses that were rated almost institutional. In other words, the normal houses provide more spatial options. This is not to say that they have more space overall, but that their spaces are better developed and that they tend to give residents the option of closing off spaces for

1-23. Rooms used for more than one function make for cozier settings.

privacy. In those houses rated most institutional there was more space that was potentially useful going unused. These spaces were either remote, restricted by staff rules, or simply not usefully furnished.

Rooms arranged to invite two or more kinds of activity result in more focused and useful spaces. This parallel use of space allows both individuals or groups to use an area without interfering with one another. In a large dining room, in addition to the table, there may be a large couch and an electric organ, allowing for several kinds of activity.

An added benefit of these arrangements is that the scale of each setting seems smaller and cozier. The living room in one house had a small seating arrangement around a coffee table on one side of the room and on the other side, a rocker and end table. Several houses made good use of existing "anchors," such as the fireplace or window, as a focus for smaller arrangements within the room.

Small furniture settings near a center of activity or traffic area are often preferred. In one home (fig. 1-10), there are two small seating areas near the open stairwell—one on the ground floor (fig. 1-24), one above it on the second. People can sit here and enjoy the comings and

1-24. Nooks in central areas that include comfortable furniture encourage group communication.

1-25. Furniture arranged against walls discourages group activity.

1-26. An institutional setting reinforces an institutional mindset.

goings of the house. In another home, a telephone, small table, and chair on a stair landing became very popular for both casual and private conversations. Communication seems easier to initiate and less forced than it would in a room with a closed door.

In the kitchen, chairs around a big table allow people to watch cooking or to keep the cooks company. People who may not actually cook can become part of the kitchen routines anyway. The table acts as a prop, providing an excuse for being there.

Furniture arranged against the walls of a room often keeps people too far apart for comfortable socializing. In many of the houses surveyed, the shape of a room does not in itself determine a useful arrangement of furniture: quite often the opposite is true. In many houses furniture was arranged against the walls of a room—like the dayhall in an institution—regardless of what traffic patterns through the room were or how the room was meant to be used. In two such houses, chairs made a circle nearly the size of the large, formal living room. Furniture arranged against the walls also creates wasted space in the center of a room and ignores the room's potential for having two or even three smaller arrangements within it.

In some of the houses surveyed, residents are part of the decision-making process; in others, they are not. Our subjective impression is that where rules are imposed unilaterally by staff and the place is arranged by someone else, the residents are intimidated and unwilling to make changes. The house in figure 1-12 was decorated by an outside organization; consequently, it looks too perfect to touch. Residents have never made a mark on it, put up a picture, painted a wall, or done anything to make it theirs. People in this house seemed passive, almost never, for instance, even closing the sliding doors between shared rooms to create a private place, even though almost all said that privacy was an unmet need. The shared spaces, though comfortable and attractive, have not grown out of the group's needs; the opportunities to personalize, take control, make choices, and govern themselves have all been missed.

In contrast, in another house (fig. 1-3) many residents were deeply involved in planning and in the actual work when the living room was recently painted and its furniture rearranged. People there seem to be more at home, surrounded by friendly clutter: games, magazines, newspapers.

Achieving a feeling of ownership of space shared by a whole group is a difficult challenge. Even in the cooperative households at the normal end of our continuum, the furniture arrangement and decoration of shared spaces is not a resolved issue. The five residents of the house in figure 1-6 have a great deal in common and actually hold joint title to the house. Yet they still expressed some discomfort about changing the ground-floor spaces, which they themselves had originally furnished, even though change seemed indicated. Some felt that the arrangement might be "precious" to one of their housemates; others were not sure whether everyone would like changing the arrangement. The group was not yet cohesive enough to find agreement on such matters. The more residents a home has and the higher the turnover rate, the more difficult this problem is to solve.

In other homes, the spaces themselves are difficult to manage, personalize, or even arrange. For example, very large spaces without any clear identity do not readily suggest use or arrangement. People seem to consider them inviolable, never thinking of breaking them up into small areas to make them more habitable.

1-27. Furniture arranged as in theaters—all facing the television (lower left)—promotes isolation.

1-28. Here again, the focus on the television does not promote shared activity.

An interesting contemporary development is the interior decoration of the urban commune. In a number of examples in the Berkeley-Oakland area visited . . . it was very noticeable that the *bedrooms*, the only private spaces of residents, were decorated in an attractive and highly personal way symbolic of the self whose space it was. The *living rooms*, the communal territory of six or eight or more different personalities, however, were only sparsely decorated, since, presumably, the problem of getting agreement on taste from a number of disparate and highly individual selves was too great to overcome. Interestingly, the more normal family house may display an opposite arrangement, with bedrooms functionally, but uninterestingly decorated, and the living room, where guests and relatives are entertained, containing the best furniture, family momentos, art purchases, photos, and so on, and representing the collective family "self."[28]

The television set generally dominates the largest, or only, available social space. It is seldom seen as one choice of activity among several.

An often heard criticism of television is that it fosters isolation, passive consumerism, and noncommunicative behavior; media supporters argue that it can offer a casual way to come together and provide a topic for conversation. No one seems to deny, though, that television can dominate whole spaces, large or small. The television set generally occupies the most privileged spot in the most central living space. It is often placed in a secure corner that could otherwise be used as a reading or conversation nook. From this secure spot, the set dominates the room, holding most residents in rapt attention. Little talking is allowed or tolerated except during commercials.

In making our survey we found that the residents of the houses rated most institutional watched television 40 percent of the time. Most of them expressed frustration over the television dominating the living room and interfering with visiting, but no one took action to alter the routine.

In some houses, viewing patterns were reminiscent of those in institutional dayrooms. Indeed, the residents of these houses, usually the houses with little constructive programming, had spent the most time in institutions.

The four houses in our survey that were rated

as least institutional either have no television or locate it in a small room or in individual bedrooms. Television watching accounted for only 16 percent of the residents' time. The problem, however, does not seem limited entirely to size or arrangement. In one house (fig. 1-16) where passive behavior seems most pronounced, the three relatively small living rooms—called dayrooms by residents—have small, domestic furniture arrangements, with coffee tables, chairs, and a couch. The living rooms do not look like institutional dayrooms, but just the same, residents sit watching television, talking little. They neither read nor engage in self-initiated activity except playing stereo tapes. Because no one has a job, this goes on most of the day, a clear example of the physical environment's inability to correct inadequate programs.

When the staff exercises exclusive control of space, residents are correspondingly passive. In houses where such rules as "no using the front parlor except for visiting" were made unilaterally by staff, residents move around cautiously, giving the impression that they are trespassing. In several of the houses surveyed, the staff managed subtly to keep residents on the second floor even though some of the ground floor social spaces were for residents. In one of these houses the best ground-floor social space, the dining room, was used to store furniture in part, while its large dining table was designated for holiday use only. Residents stayed upstairs, mostly in two sitting rooms. In another house, the operator presides over the kitchen, and to get to the other ground-floor social spaces, residents must pass through the kitchen under her scrutiny. The front parlor is said by some residents to be off-limits, even though the operator and other residents say otherwise. In this house, most residents stay upstairs most of the time, treading lightly downstairs, even during the meals served to them by the operator.

In the house in figure 1-11, the staff controls three of the five available ground-floor spaces. A secretary and copy machine effectively occupy one forth of the area, and the staff limited the use of the kitchen, making that, too, their territory. Residents gather in the remaining two spaces at close quarters.

28. Clare Cooper, "The House as Symbol of the Self," in *Designing for Human Behavior: Architecture and the Behavior Sciences*, edited by Jon Lang (Stroudsburg, PA: Dowden, Hutchinson and Ross, 1974), pp. 130–46.

Frequently, people in group homes do not become an integral part of a residential neighborhood. While their access to community supports and facilities is very important to them, they did not attempt to maintain close contact with neighbors. There seems, however, to be a larger proportion of boarding-care homes integrated into a neighborhood than agency-operated homes. A probable explanation for this is that operators who own their own homes were originally themselves part of the neighborhood.

Agencies tend to locate houses on the edge of a neighborhood facing a main street, in a semi-institutional or semicommercial part of town, or, alternatively, in a semirural place. This may be because these locations present little problem of acceptance and because houses are more often found available there than in socially tight neighborhoods.

Kitchens and Dining Rooms

In each house we surveyed, we made a special effort to observe the evening meal. The kitchen and meals are central to almost any community residential program. At its best, the group meal is also the time when everyone comes together to share the experiences of the day, to rub elbows, or to lend a sympathetic ear. The ritual of the "family-style" meal is seen as important in stimulating rapport among residents and staff.

Having a daily group meal was a goal of almost all of the houses we visited. However, simply having the meal does not guarantee mutual support and communication. In the institutional-type homes, we found that little time was actually spent at the meal. People tended to "eat and run" without lingering over coffee. One reason for this may be that eating was the only part of the ritual the residents participated in. By being excluded from planning the menu, grocery shopping, and preparing the meal, the residents were narrowly defined as passive consumers of the meal.

1-29. The kitchen can be the ideal informal gathering place.

In most cases, one person was responsible for all the cooking. Not surprisingly, in these houses, the kitchen was considered the domain of the cook, and the order and style of the room reflected the cook's personality. The residents were actually considered in the way. Typically these houses had more rules and restrictions about kitchen use. In some cases the kitchen could be used only when meals were not being prepared. In others, the kitchen was locked when the cook was out. In one such house, although the residents are perfectly competent, they were not allowed to use the kitchen at all.

Obviously it can be argued that having one cook prepare all meals is more efficient and practical, but this method is institutional in character and leaves control in the hands of the staff. The residents are excluded both from access to an important room in the house and from access to useful and necessary skills and activity. It is a vicious circle: the residents are not considered competent enough to use the kitchen, and they are denied the opportunity to practice skills to increase their competence. The message to the residents is that they must continue to be dependent on others.

1-30. When the kitchen becomes only the cook's domain its potential as a space for socializing is lost.

1-31. A breakfast nook can be an inviting group space.

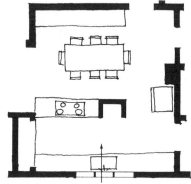

Figure 1–31 photographed from here.

1-32. Layout of kitchen shown in figure 1-31.

In the more normal group-home settings, meals became not only life-sustaining, but life-sharing as well. Residents talked and joked and enjoyed one another's company. They helped in meal preparation, and the meals were more lively. As a whole, these houses seemed to have a strong sense of group togetherness. Taking turns helping to prepare meals provided an opportunity for everyone to take part in meeting the group's needs. Residents felt more responsible, more a part of things and, hence, less passive.

Planning menus, grocery shopping, cooking, and cleaning up afterwards are all necessary to day-to-day life. Acquiring these skills is an important part of the transition from the institution to the outside community. The act of planning and preparing a meal also builds self-esteem and confidence.

In places where mealtime participation is shared, rotating responsibilities requires some programmatic effort. There must be a schedule and someone to coordinate groceries and be sure of supplies. However, the rewards for the extra effort make it well worthwhile.

In several of the houses, a breakfast nook, booth, or small table with chairs in the kitchen became additional social space even when meals were not being prepared. Flexibility and variety of choice were increased. People could congre-

gate comfortably without being in the way. They could talk, observe, lend a hand, have a late night snack or take a coffee break. The other rooms were not as crowded, and people did not seem to be on top of one another all the time. Because the atmosphere in a kitchen is often much more relaxed than in other rooms, people are not on "display" and actually talk more freely.

In some houses, the kitchen is an alternative to the formal dining room, and in most cases, became the hub of activity for the house. It is an active and comfortable place, with people coming

1-33. Booths encourage group gatherings.

1-35. A small table with chairs in addition to a large kitchen table provides a choice of social spaces.

Figure 1–33 photographed from here.

1-34. Layout of kitchen shown in figure 1-33.

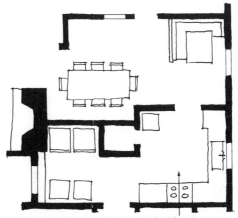

Figure 1–35 photographed from here.

1-36. Layout of kitchen shown in figure 1-35.

and going, exchanging words, and finding out about the day's events.

In some houses, the dining room is an actively used social space at all times of the day; in others it is used at mealtimes only, and conversely eating in other rooms—snacking in front of the television for example—is discouraged. In some houses surveyed, a piano, couch, or record player in the dining room encouraged use at nonmeal times. In one house, the dining room is the site of an ongoing game of gin rummy.

The narrow interpretation of room use seems a waste of potential and limits the alternatives for a variety of activities and relationships. If a resident's only choice is between being alone in a

bedroom or being with the whole group in the living room, then opening the dining room for other activities is an easy way to increase the options available.

The traditional dining table, though symbolic in the sense that everyone could gather around it, does not work well with a large group for good interchange. The traditional shape itself—long and rectangular—also inhibits a natural flow of conversation. Most of the homes in the study had just such a large table placed in the middle of the dining room. When five to eight people used it, there seemed to be no problem with conversation. But in larger groups, with larger tables, the distances between people made con-

1-37. An inviting kitchen can become a hub of activity.

versation difficult. In these cases, the meal was treated as a more formal situation, and the possibility of lively everyday conversation seemed to be stifled.

Two of the homes with large groups, one of fifteen members, the other twenty-two, arranged their dining rooms with several small round tables, four feet in diameter, instead of one large one. This arrangement worked well. Residents could choose where and with whom to sit, and conversation was animated.

Bedrooms

Bedrooms ought to be more than just a room with a bed. Especially in group homes where there are no options for privacy in shared spaces and bedrooms are also shared, the bedroom must satisfy more needs. The stereotype of just a bed and dresser implies that only sleeping and dressing happen there. But these rooms are also private spaces for thinking and dreaming, for being alone or for being intimately alone with another. Books are read there, letters are written, music is listened to.

On the whole, people in the houses surveyed reported spending much of their free time in their bedrooms—second only to the living room. They also considered their bedrooms to be their favor-

ite room in the house. Perhaps this is because in the bedroom, they can, at least to some degree, control what happens and how the room looks, decorating and moving furniture to suit themselves. In one house the bedrooms were so barren of personal possessions that they appeared to be unused (fig. 1-38). The people seemed more like guests than residents, and the staff did little to dispel this feeling. Without staff encouragement even residents with private bedrooms often do not use them as sanctuaries or personalize them. Although more than 75 percent of the group-home residents interviewed preferred a single to a shared room, those residents with singles often did not personalize or take control. Again such activities depend on the attitudes of the staff.

If bedrooms are subject to intrusions by staff, residents are not likely to feel that the room really belongs to them. Paternalistic ideas and regulations about order and cleanliness, characteristic of institutions, are another sort of intrusion, destroying feelings of ownership and responsibility.

In several houses surveyed there were double rooms set up as if they were to be occupied by identical twins, with identical beds, dressers, and bedspreads. While this may make it easy to purchase furnishings, it says little about recognizing each person's individual tastes.

In other homes, the program itself encourages residents to take over their rooms, to personalize, rearrange furniture, and even repaint. At the same time a balance between being alone and spending time with others is also encouraged.

Staff members sometimes view shared rooms as a way to keep residents out of trouble and prevent them from isolating themselves. Yet finding privacy and opportunities to be alone in shared bedrooms is a difficulty frequently mentioned by residents. Several home operators in the survey preferred having double bedrooms, saying that they foster peer support. Compatible roommates can be very supportive of each other, but if such a system of double rooms is forced, residents may simply find themselves required to share their lives with an incompatible person.

Part of the problem with shared bedrooms is that there is little effort to define and separate

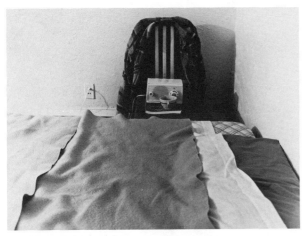

1-38. A bedroom devoid of personal possessions discourages a sense of ownership.

1-39. Some programs encourage residents to personalize their bedrooms.

the space into two distinct parts that can be used independently. In some houses, this is less of a problem because there are other areas to which residents can retreat—a study, sun porch, or window alcove. The room itself is often too small to give residents minimal space and privacy. Such bedrooms can lead to tensions between otherwise compatible roommates. In some cases minimum codes and standards governing group homes are responsible for skimpy rooms, allowing 60 square feet per person. There is just no room to divide and delineate a space this minimal once the beds, dressers, chairs, and any other furnishings are set up. There is barely enough space for people.

The average space for each resident in the houses surveyed is 130 square feet. This ranges from a low of 63 square feet per person in a house with dormitory-style bedrooms to more than 200 square feet in a house with nothing but single rooms. However, both of these figures are above the minimum standards. Such standards deal only with the quantity of space and not with quality. The shape of the room, location of windows and doors, arrangement of furniture, are all factors that determine quality and comfort for the inhabitants. Furniture arrangements can give each person a little more privacy and defined territory.

In dormitory rooms, with three to six people,

1-40. A bedroom should be personalized to accommodate more activities than sleeping.

it is most difficult to provide for personal space. In double rooms schedules can usually be arranged so that each person can have the room alone from time to time. In dormitory rooms, this is almost impossible. People in dorms rarely have privacy or the luxury of being alone in their own room. Of those we interviewed, most complained about differences in habits and routines— times of awakening and retiring, lights, alarm clocks, radios, and interrupted sleep.

In one home for adolescents, the occupants of

1-41. Dormitory-style rooms can become impersonal wall-less territories.

1-43. Shared rooms make privacy almost impossible.

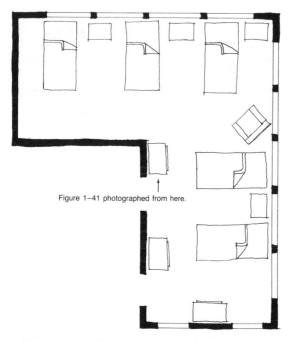

Figure 1–41 photographed from here.

1-42. Layout of bedroom shown in figure 1-41.

a dormitory have a "pecking order," with several girls in leadership positions pressuring the others about everything from room apportioning to dress standards. This seems to be a way of dealing with the dormitory's inherent lack of privacy and territory. It also reinforces a dominance based on power that conflicts with program goals.

Dormitory-style rooms generally allow the least square footage per person. Little effort is usually made to define individual territories or provide privacy. In some cases each person has a corner of the room with a bed and dresser and little else. Bureaus, dressers, and freestanding closets can all be used to divide rooms into separate areas without building partition walls, but, in the survey, these attempts were generally failures.

Bathrooms

Too often, bathrooms are thought of in no more than functional terms, but they are much more than just a place for physical hygiene. Primping and preening, putting on makeup, showering or bathing—all make a difference in image and self-image. Bathrooms are absolutely private places in which people can be themselves with abandon.

In the survey houses visited, the bathrooms

were tight, basically functional, no-frills spaces. The number of fixtures available to residents generally exceeded current licensing standards. Despite this, about 40 percent of those interviewed reported that they occasionally have to wait to use the bathroom. This is not surprising for the licensure standards are absolute minimums. Typical current standards require one toilet and one lavatory for every six residents and one tub or shower for eight residents. The standards also require "adequate and private" facilities with hot and cold running water. The houses we surveyed averaged one toilet and lavatory for four residents and one tub or shower for five residents.

Several of the homes have remodeled and compartmented bathrooms for more efficient use. Compartmenting means that different fixtures are behind separate doors, allowing more than one person to use the space if necessary. While remodeling, a second sink was often added—either side by side or back to back—which also increased the capacity for use (fig. 1-45). Two houses have an extra shower in the basement, easing the burden on upstairs facilities. This is especially important if employed residents shower after work. One house—a former convent—has sinks in every bedroom. While this is very convenient, it would be a costly addition.

By locating a half bath on the first floor, near the most public areas of the house, staff members rarely invaded the residents' bathrooms upstairs. It is also a convenience for residents and their guests. Increasingly, with new attention to access for disabled people, first-floor bathrooms are required, especially if federal funds are involved in remodeling efforts.

Some state and federal regulations discourage storage of personal items in bathrooms. Unfortunately, by applying this as an absolute rather than optional rule, group-home staffs have made bathrooms appear just a bit less than normal. They seem rather cold, stark, and impersonal without the presence of toothbrushes, extra towels, hair dryers, and so on. Most residents carry their toilet articles back and forth from their rooms. A few homes have large linen closets near the bathroom, providing easy access for towels, wash cloths, tissues, toilet paper, soap, and other supplies.

1-44. Bathrooms should be more than just rooms with plumbing fixtures.

1-45. A bathroom with two sinks cuts waiting time.

In most houses people are expected to clean the bathroom after using it. In addition, a once-a-week cleaning is usually an assigned chore, rotated among the residents. In most of the survey houses, bathrooms were in presentable shape, though cleaning and maintenance is always a problem in places where a group of people live together.

Laundries

Laundry day for most of us is something to be endured. Yet for some, the laundromat is a place to meet friends, do some reading or mending, or simply hang out.

Clean neat clothes often reflect a person's feeling of self-worth. Learning to care for and keep track of one's clothing and linen is a way to develop pride in personal appearance, feelings of competence and control of personal possessions. It is essential to self-sufficiency.

Laundry is clearly one of those everyday life necessities that a resident in a group home can learn to deal with, yet it is a missed opportunity in many group homes. It seems ironic that in those group homes with only one person on staff, the burden of doing everyone's laundry is usually assumed by that one person.

In the boarding-type homes, daily routines seem often to resemble institutional life. The operator tends to take few risks—not allowing residents to participate in activities such as laundry or cooking. This may be because the operator owns the house and any risks involve property as well as reputation, or because the operator has fallen into the role of parent to the residents.

In contrast, in homes with several staff members and intensive programming, residents usually do their own laundry, taking advantage of some of the possibilities suggested by this activity.

No matter what laundry regimen is followed, group homes rarely make their laundry place attractive and supportive. In all but two out of fifteen group homes we studied, the washers and dryers were in the basement, usually without adequate areas for folding clothes, storing supplies, ironing, or sewing (fig. 1-46). There was

often no seating and apparently little realization that games, television, or other potential activities might be made available as pastimes while waiting. One place did have a pool table nearby.

Another, a group home for adolescent girls (fig. 1-8), came closest to integrating the laundry with the rest of the house. Located on the ground floor next to the kitchen area (fig. 1-47), it seems to have become a gathering place of sorts, for it is heavily used by the girls. The houseparents reported that the most intensely used route is from the living room to the laundry.

1-46. Most laundry areas are poorly planned basement spaces.

1-47. Laundry rooms can promote group activity, as this space does.

Staff Spaces

All of the homes we visited, except for the cooperative households, have some staff, varying from parttime to fulltime live-in. Some staffing patterns are dictated by funding criteria or licensure standards; others are a matter of preference. The staff fulfills many different roles—parent, administrator, cook, counselor, housekeeper, friend. To function well and do their jobs effectively, one of their basic requirements is physical space—places to work, places to be alone or with others, and places of their own.

On the whole, we found that the staff tended to control most of the shared space in the house. They established rules about space use and activities, furniture arrangements, and tidiness. In many cases this is built into the staff role, but it means that residents have less opportunity to make decisions and to exercise responsibility.

Live-in staff members, hired by an agency to run a program, are more likely to share control of the house with residents than are operators who own the house. The homeowner takes a proprietary interest in controlling what happens there. In such houses, the residents appeared to us to be boarders rather than participants in group living.

In one case, the owner-operator lived elsewhere but spent most of the day at the group home, on the first floor, cleaning and preparing meals, thus effectively discouraging residents from using the first-floor space. In another owner-operator house, the owner used the entire first floor—except for a private suite and the kitchen—to store excess furniture, rendering the space unusable by the residents.

In contrast, in two other surveyed houses, the live-in staff were not owners but agency employees. In both these houses, their use and control of shared spaces was more sensitive to residents' needs. Other agency staff members also played an active role in the program and visited the house regularly.

Whether for owners or hired staff, adequate personal space for living in is important. This should include bedrooms, bathrooms, and living space. Being "on-duty" twenty-four hours a day, every day, can be a tremendous strain. For the staff to be able to get away for a while requires a defined and separate living space.

Staff control of resident spaces may be explained by their individual personality, their concept of their role, the type of program they are involved in, or a combination of all three. For staff who work shifts, having adequate, well-defined offices makes it less likely that they will use and control shared resident spaces. Without offices, staff members must use the shared spaces of the house, and simply by spending time there, they can convey a message that it is staff territory, possibly inhibiting use by the residents.

Sometimes, even though staff have adequate office space, they still tend to dominate the rest of the house. In one kind of program, the staff decides unilaterally how various rooms will be used. The rationale is that the staff provides an element of continuity in a transitional program where the average resident's stay is from three to six months. Most programs set up in this way concentrate on educational or vocational goals, with residents out of the house for eight hours of the day instead of spending the whole day in the house. Even so, when the staff makes the decisions and rules for residents, the mode of operation becomes institutional.

The location of staff space can influence the amount of control they exercise and can communicate their attitude toward residents. Several of the homes have a staff station or reception area at the front door, which would seem more appropriate to an office than a home. The arrangement gives a clear message about who is in charge and is an effective means of keeping track of the comings and goings of everyone in the house. The halfway house for prisoners (fig. 1-7) has no office at all in the house and deals with people's comings and goings as part of the issue of trust and responsibility.

In several houses the room selected for the staff office is centrally located or near the major shared spaces. In this way, the staff stays on top of things. In contrast, in other homes staff offices are at the back of the house, upstairs, or even in a carriage house. These locations clearly reflect the need for a private workspace fitting the functions of an office rather than a kind of sentry box.

Another problem is that often the rooms selected for staff offices are the most desirable spaces in the house, those small private rooms that could be closed off comfortably for two or three people to use at a time. When these become staff rooms, the range of spatial options for residents is more limited. Larger public rooms—living room, dining room, kitchen—may be the only spaces left. In houses with no single bedrooms, small closable rooms are especially valuable for expanding choices for privacy. Being able to close a door is a sign of trust that carries with it privilege and responsibility.

In two of the houses a compromise has been agreed upon and it seems to work well. The office is left open for residents to use in the evenings for writing letters, making personal phone calls, or studying. Confidential material—the eternal excuse for keeping residents out of some areas—is easily locked in a file cabinet. To residents, access to the office space is a sign of trust.

Special Features

Some features in old-house construction are especially useful in group living. Nooks or alcoves can help residents to organize space in their own fashion. A bedroom alcove might

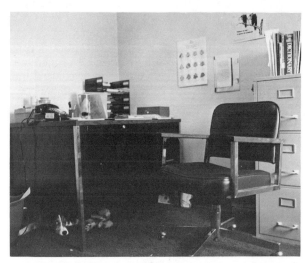

1-48. Staff space should not encroach on residents' space and ideally should be open to all—even pets!

become a bed nook, a study nook, television corner, or even a dressing room. The space suggests the use differently to the people who fill it. In figure 1-49, the dormer became a bed nook, and the added small spaces, corners, and wall surfaces lent themselves to personalization.

A small space just off a heavily used, shared area can give a sense of privacy without isolation. Comfortable for one or two people, these spaces often become favorite spots.

Indentations in walls lend themselves to storage as in figure 1-50, in which the small leftover spaces adjacent to fireplaces have been put to good use.

Window seats have universal appeal. In institutions or new buildings, there are few casual surfaces such as the one in figure 1-51.

In newer housing, corridors are mainly fire escape routes, but in older houses, hallways are large enough for bookshelves, record players, chairs, phones, and other useful furnishings. Front stairs make good spectator seats, and back stairs provide a space for private talks. A wide hallway outside a bedroom as in figure 1-52 can be a valuable shared space. It can host a furniture grouping or be a good storage space. By becoming a place shared by the group of people whose bedrooms open onto it, it turns the floor into a suite.

Open stairwells are invitations to lively communication. In figure 1-53, there are places to sit at the top and bottom (see fig. 1-24) of a lovely old staircase, and people also lean on the balustrade to talk to people below.

Pictured in figure 1-54 is a small low hallway under a stairway. It separates the breakfast nook from the rest of the house (see fig. 1-55), creating a behind-the-scenes area.

Porches extend a house's living space and increase the choices of shared spaces available in a house. They provide a way to remain securely on one's own territory while watching the outside world go by. Front porches seem to win the popularity award probably for this reason.

The floor plan in figure 1-2 is one of two houses side by side, with kitchen and dining spaces for all residents in one house. Since our survey was made, the houses have been joined by an enclosed passageway. Previously, the house without the eating areas was usually rather dead. It

1-49. A dormer becomes a sleeping space that encourages personalization.

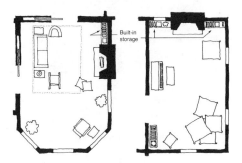

Built-in storage

1-50. Leftover space can be used as storage space.

1-51. Window seats can be used as gathering places, knickknack holders, or reading nooks.

may have been simply easier for residents to stay in the house where meals were prepared rather than walk back and forth, or most residents may have found it pleasant to be near the center of activity. The dead house did seem a lonely place, used occasionally as a place for residents to retreat from the crowd or staff. After the remodeling, this pattern of use changed, so that now there are smaller groups, actively using both houses.

Another home (fig. 1-7) had a carriage house at the rear of the lot that housed an office and several bedrooms. The arrangement was very suc-

1-52. Large hallways can serve many shared purposes.

1-53. Open stairwells promote informal conversation.

1-54. A small passageway to a single area lures residents to the private space beyond.

cessful, providing a good separate area for staff. Their working style created an open-office atmosphere attractive to the residents who often visited the staff or other residents whose bedrooms were in the carriage house.

Storage space was sorely lacking in most houses surveyed. Some homes lacked adequate storage space for kitchen staples; others were without drawers, shelves, or niches for games, paper, pencils, or other supplies. By and large, the older homes provided ample built-in storage and cupboard space. Such houses usually include pantries which not only provide storage space but also help to separate kitchen work areas from the serving and cleaning-up spaces. In the house in figure 1-12, the pantry provides a low, winding transition space separating the private breakfast nook from the rest of the house. Unfortunately, in more than one house, food has

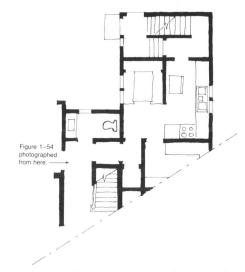

Figure 1-54 photographed from here. →

1-55. Layout of the passageway shown in figure 1-54.

1-56. A front porch works as an appealing shared space.

1-57. This old-fashioned porch swing allows residents exposure to the outside world in the security of their own territory.

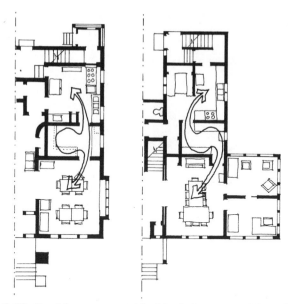

1-58. Pantries can provide ideal storage and transition space.

1-59. Basements are often used to store all sorts of things, including nonperishable foods.

been pilfered from the pantry, necessitating locking it. As a result, residents have no evening access to the kitchen.

The residents in one house (fig. 1-10) do their own canning, using their garden produce and storing it in the basement. In other houses, too, all sorts of nonperishables are kept in basement rooms (fig. 1-59).

1-60. Sliding doors offer a variety of space options.

Doors between shared spaces increase space options. Particularly in older homes, we found built-in doors available to close off adjoining rooms. In some cases these are heavy sliding doors, and in others French doors or "bi-folds" separating sun rooms or libraries from other areas (fig. 1-60). In homes with ten or more residents, the sliding doors of the living rooms are seldom used, perhaps because one person feels uneasy making such a change for everyone. Smaller rooms are more often closed off in those houses. These available alternatives, even if used only once in a while, make group homes more habitable.

Cleaning and Maintenance

Cleaning and maintenance are bound up with feelings of ownership. In those houses surveyed where the residents seemed to feel a greater sense of ownership and control, cleaning was less of a problem. Without these feelings, there is very little incentive to see that things are kept tidy, clean, and comfortable. An overly tidy place can feel uncomfortable, limiting the spontaneity and range of casual activity. An unsanitary place is worse, however, for not only is it uncomfortable but it can also be closed down. In the houses surveyed, we found that just a little clutter is a reliable sign that people feel at ease.

A weekly cleaning day involving everyone can become a community event with a purpose. For some groups, designating one day a week—Saturday, perhaps—seems to be an efficient and effective way of keeping house. In addition, a continual periodic maintenance program to survey, repair, and replace items throughout the house is beneficial but usually overlooked. Typically, in the homes surveyed, much money, time, and effort went into setting the house up. After the first year or so, things began to show signs of wear and tear. Instead of anticipating breakdowns, however, most house groups waited for the inevitable crisis before acting. Private operators, in particular, say that they did not anticipate how high the costs of maintenance would be.

A good rule of thumb is to figure that almost everything must be replaced every twenty years and to set aside 5 percent of the value of the house accordingly, each year, for maintenance and repair. Having a mixture of new and old furnishings and equipment helps to ensure that everything will not fall apart at the same time. This also gives the positive impression of a place continually being renewed.

Building Codes and Zoning Ordinances

Most of the homes visited have confronted —but not resolved—the conflict between building safety and the desire for a noninstitutional environment. Much of the institutional quality in the survey houses resulted from compliance with building and fire safety codes: exposed sprinkler heads and pipes; wired-glass vision panels, panic bars, automatic closers, and resurfacing on doors; lighted exit signs; walled-in stairwells; large fire alarms and battery-powered emergency lighting packs; commercial kitchen fixtures and posted foodservice licenses. Each of these items conveys a powerful institutional message. Obviously, safety from fire is extraordinarily important for every single space, but this can usually be achieved with sensitivity both in terms of equipment and program, without destroying the residential character of the house.

Code requirements relating to safety and habitability originate from several government or accreditation agencies. Codes vary from one place to another, so, for a particular house, investigation is necessary to know what requirements apply. These requirements may vary from do-it-yourself changes, such as posting evacuation route diagrams and conducting fire drills, to major structural alterations such as enclosing stairwells. Earnest work with inspectors, some creativity, extensive examination of alternatives, and serious safety consciousness can usually produce changes that keep the house safe, intact, and still residential. It is true, however, that codes are sometimes misused purposely to discourage the establishment of group homes; when this happens little can be done.

Exit requirements are at the heart of codes dealing with fire safety. The central principle is that there must always be two separate ways out so that, if one is blocked by fire or smoke, the other may be used to escape. This is a problem in a house with bedrooms on the second floor where windows cannot be used as escape routes (as they usually can be used on the first floor). The typical problem is that there is only a single stair or that there is an open stairway (which would not be considered "separate," since smoke from below could easily fill the escape routes above).

Trade-offs are possible here for sometimes there is a connection between the number or type of exits and requirements for sprinklers; fewer or more open exits are considered adequate if there is a sprinkler system. One of the homes surveyed had chosen to install sprinklers for $15,000 rather than wall off their central, open staircase. While the costs of the two alternatives were similar, walling off the main stairs would have meant making the house darker, destroying the most striking, central and well-preserved feature of the house. It also would have blocked the lively interactions between people that now take place up and down the open stairwell.

There are other alternatives to blocking off open stairways. One of the survey homes, for example, used an external stair connected to an existing first floor roof. In most of the homes, however, examples of institutional character and institutional messages intruded into the house, destroying some of its residential quality.

AN ENVIRONMENTAL
APPROACH

Approaching group homes from an environmental perspective is an underlying theme of this book, although one that is difficult to define. Essentially, every social situation is set against a physical space that has its own geometry of relationships. An environmental perspective allows us to see the geometry of relationships, to evaluate it and design for it.

The geometry of relationships consists of the associations and distances between people as established by the physical environment. The layout of space and the rules that are made about how the space is used affect peoples' behavior toward the space and each other. Evaluating the geometry that results means determining whether or not it serves the needs of people within the situation. From this perspective, the aim of design is the creation of an environment that satisfies these needs by allowing for an appropriate geometry of relationships. Questions immediately arise about whose needs are being served, and who controls the process.

Figures 2-1 and 2-2 show the same picture twice, with the environmental props painted out of Fig. 2-2. Even without the environmental props, the associations between people remain obvious (customers and clerk), certain distances between people are being maintained (close enough for conversation but no touching), and certain rules are being observed (select what you

want to buy with reasonable quickness and make room for the next customer; do not go behind the counter). The environment is quite simple—counter, display lighting, canopy, raised platform—yet it is powerful in structuring the geometry of relationships between people. Imagining people floating in a void of empty space is a useful way of seeing how social situations are set against physical space. This geometry of relationships suggests the environment and reveals its importance to the situation.

Recognizing the relevance of specific spaces to behavior is the beginning of an environmental approach to design. This perspective offers new possibilities for design solutions.

Supportive Settings

There are special places that people maintain better or feel livelier and happier in or seem more at home, more themselves, in than other spaces. Most people have memories of at least one such place. For some it might be a grandmother's kitchen, remembered for its pungent smells and sweet associations with a special, loved person. For other people a well-remembered street corner where they waited for a bus might be such a place.

2-1. The physical environment often establishes social relationships.

2-2. Imagining a space without environmental props enables the designer to see relationships clearly.

2-3. A supportive space fosters learning.

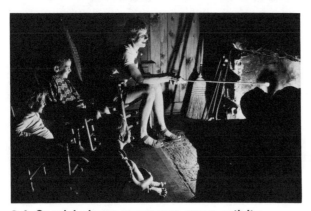

2-4. Special places encourage group activity.

Or it might have been a chemistry classroom with intriguing glass flasks and bottles of chemicals lining the shelves, a place with an order and organization that was clearly understood.

All of these special places share a common thread. Each acts as a supportive setting, as a kind of stage. They have some meaning or purpose or life of their own that is appealing, and these places change over time to reflect moods, to comfort or surprise their users. Most people who treasure such remembered places also try to re-create them.

The fabric of such places is woven of ordinary substance—durable, practical materials; decent daylight and ventilation; straightforward friendly objects; color used with care. The scale of the space and furnishings are proportioned for people. It is a flexible space with room for supplies and ornaments, reflecting the personality of the user.

Form and function are inseparable in such places. Not only are they visually handsome, whether streamlined or filled with clutter, they also invite involvement. Such involvement allows people to grow, to become more whole. It is this potential that distinguishes supportive settings.

Though everyone benefits from supportive settings, people who live in human service places such as group homes are especially affected. By living in a setting that supports the rich range of behaviors, a person can move from an institutionalized, passive, dependent life toward an active life of participation, and choice.

2-5. Although ordinary, a supportive setting can encourage participation and choice.

2-6. A special space is comforting—a place one returns to for sustenance.

Trust

Mistrust and suspicion affect places in most unpleasant ways. From the video cameras that monitor activity in department stores to the armed guards in fast-food restaurants and other places, the environment reflects the fears of those who control it and use it.

A physical environment that is structured on mistrust is especially damaging to those who use it. By being set up to anticipate the worst behavior, it relays a message to every person—trustworthy or not—within that environment: the worst is expected. In a mental hospital, nursing stations are always located to make it possible to survey the wards, reinforcing the message that wards must be watched constantly. Private areas are eliminated for people might hide or fight or have sexual contact in them. The hospital is escape-proof and vandal-proof. Carpets are removed because people might burn them with a cigarette or urinate on them. A new employee or a patient coming into such an environment reads the message immediately. Ironically, some of these measures cause more problems than they solve. Denying privacy will generally result in more fighting, yet privacy is denied with the idea

that fighting will be prevented.

Of the group homes surveyed, those on the institutional end of the continuum reflected this lack of trust. There were several ways in which mistrust surfaced. One was the way in which the staff dealt with residents—barking orders rather than addressing fellows. In other cases the use of private space, even bedrooms, was restricted. This was not consistent but seemed to depend on beliefs about the risks involved. In some houses single bedrooms were consciously avoided so that residents could keep tabs on one another—limiting the use of drugs or alcohol, reducing the risk of suicide. Mistrust was also apparent when residents reported that the staff was not understanding enough or invaded their privacy.

Such staff behavior may be motivated more by real concern than simple malevolence. For example, for a resident to be alone with a visitor may seem risky to some members of staff. There are problems of legal responsibility where the resident's physical safety is concerned. On the other hand, however, residents have a real need for private inviolable space.

It is too easy to develop environments where people are considered a threat and restrictive measures are taken; to avoid a drift in this direction takes conscious effort, involving a combination of mutual respect, communication, and a willingness to take risks.

Trust is also conveyed or withheld through rules and decision-making procedures. Rules in themselves are not always resented, but if they are implemented unilaterally they may cause mistrust. It takes trust for staff members to release some of their hold on the decision-making process, but it develops trustworthiness in the residents. In this way, trust has everything to do with whether or not residents feel that the house is really theirs to use.

Control and Ownership

Control, as we mean it, is having a voice in making decisions about the environment. Ownership, closely allied with control, is not the legal ownership of having title to a property but the feeling that develops among people who spend a lot of time in a place and have some measure of control over it. Such feelings are discouraged in our society unless people literally own a place. In spite of this, they do occasionally develop. When tenants have together successfully battled the landlord for better maintenance, they feel that the building is their own. Such proprietary feelings can develop on whole neighborhood blocks—particularly ones with front porches, which naturally give residents a chance to keep an eye on the streetscape.

That residents feel control and ownership is necessary for a successful group home. Each individual becomes a part of the house picture.

Having real control of one's own room is a first step, but this does not happen automatically in a group home. There may be no better way to establish this attitude than for each new resident to start out with an underfurnished room and be allowed to make some choices about what furniture will go into it.

Control and ownership are more difficult to achieve with shared spaces than with private spaces. Developing the right attitude about the physical setting is important. If everyone believes that the formal furniture arrangement and the color-coordinated interior decoration are permanent, participation in environmental decisions will be limited, and feelings of ownership unlikely.

Part of what must be understood in shared spaces is the tension between personal and group style. Regularly scheduled group meetings to plan changes or to work out problems automatically relay the message that control is being shared. Group influence may imply compromise and limitations on individual style, but at least it prevents an impersonal and barren atmosphere from developing.

Places people use become theirs. The more residents are encouraged to manipulate and use the house and its facilities—from such simple tasks as closing doors and straightening books to working on cars in the driveway—the more ownership they will feel.

Privacy

For many people the concept of privacy means being quiet and alone, but there is much more to it. For someone who has been institutionalized, privacy may be as simple as bathing and dressing without being watched. For others, it does mean being alone, or with one other person or a group, conditional always on

2-7. Proprietary feelings—a sense of control and ownership—are important to successful group homes.

2-8. Residents should feel free to make use of a home to support their needs.

choice of company and behavior and conversation.

Being able to be private usually depends on the physical setting. Since privacy means control over contact with others, the physical place must give us control over what we let others hear or see about ourselves. Walls, partitions, visual barriers, physical distance, doors, and sound-absorbing materials help to attain privacy. Because people spend a great deal of time in a house, it should have a variety of such options for privacy ranging from large spaces for groups to small areas for individuals.

Some privacy is essential to normal behavior. Without it people feel crowded: they may withdraw and set up psychological barriers. Some people set up other interpersonal barriers such as exclusive cliques—a cruel "pecking order" based on physical dominance replacing walls and doors. Also, private acts, when forced into public view, become "inappropriate behavior."

Many social conventions surround privacy, and most people know and respect them. Knocking on a door before entering a room or allowing two people exclusive use of a restaurant booth are courtesies that are best learned in a house where privacy is practiced in all its forms.

2-9. Privacy is essential to *every* person.

Crowding and Concentration

The requirements of minimum square footage per resident are basic to group-home licensure and funding. The standards are set up to prevent crowding and are comforting partly because they seem so precise and objective. Almost every state has licensure rules that require either 60 or 65 square feet for each person in a shared bedroom and a large amount—typically 80—in a single bedroom. In the shared spaces of the house, the required square footage ranges from 60 to 75 square feet. These rules seem to have come from district federal court orders of the early 1970s concerning right-to-treatment and "least restrictive" environments in institutions. Although these standards may seem neutral, there are several reasons why adopting these figures as general standards has proven harmful.

Minimum standards have a way of automatically becoming the *maximum* standard. Since the standard requires less space for each person in shared bedrooms, it motivates operators to provide no single bedrooms. This was certainly not the original intention of the requirement. However the concept itself is faulty, for where space is shared, each person actually needs more area than in single bedrooms. And by emphasizing square footage, many other qualities that make a bedroom livable are ignored. The result is often a standardized, rigid, nursing home mentality. For twelve residents, six bedrooms are often arranged in a row, each one 10 feet wide and 12 feet long, in fulfillment of minimum requirements. In addition one standard occasionally combines with another to limit options. For example most standards exclude the use of day beds, yet this is a simple way to increase space use.

The problem with all such standards is not so much whether a specific requirement is adequate but whether the minimum standards should be used as a design guide. However, it is important to recognize that these figures *are* extraordinarily small. In making our surveys, moreover, we noticed that the square footage per person needed to make a comfortable house varies with the actual number of residents—fewer square feet *per person* are necessary with fifteen people, more with five people.

We approached the problem of crowding subjectively, beginning with our feelings of concentration in the homes surveyed. After calculating the available shared space per resident and observing levels of comfort, we arrived at some rough estimates of workable figures. The lowest house had 69 square feet and the next two lowest, 90 square feet of shared area per resident. The former is, without a doubt, overcrowded, stuffy, noisy, and oppressive. One of the two houses with about 90 square feet per resident did not seem too small, but because it had twelve residents, the total square footage of the house was quite large. In addition, residents had widely varying schedules, so that not everyone was home at once. In the second house that had about 90 square feet per resident, while only eight people lived there, they all had similar schedules, and the house did seem crowded.

On the other hand, some homes seem to have too much vacant, unused space. Some shared spaces, by their location or the rules for their use, contribute to the square footage figure of the house but not to the life of the home. Bedrooms for example are sometimes impersonal and uninviting.

Other factors in crowding may be the number of rooms, their arrangement, circulation paths, accessibility of space, rules about the use of space, means for separating one room from another, the relationship of a place to the outdoors, the type of yard, the number of residents, and even their schedules. In any event, shared-space figures should include only the rooms available to residents for their free use, not those reserved for special occasions or for visitors only.

In the group homes surveyed, acoustics played a role in perceptions of crowding—how sound carried through the house; whether doors could close off spaces as needed; how much distance lay between noisy and quiet areas. Ceiling height also seems to relate to crowding, as does the flexibility of the space. Separate activity "pockets" in shared spaces result in several choices for groups to congregate, rather than only one large space for one large group. Outdoor spaces or nearby community facilities can also reduce perceptions of crowding by giving people more choices of places to spend time.

All of these factors affect concentration of people. Square footage per person indicates clearly the overall density of a place, but if everyone occupies only one area, the perception of crowding changes accordingly because people are concentrated.

In the survey house in figure 1-9, a former college dormitory has long "double-loaded" corridors of bedrooms (rooms off both sides), with cavernous lounge spaces at one end. Of the homes in the survey, this one has the second largest amount of shared space per resident, but it feels crowded because many residents can usually be found in the interconnecting lounges, all concentrated within sight and earshot of each other. Also, most movement is through the corridors, forcing residents to pass one another frequently at close quarters.

These reactions to the dormitory were subjective. A more carefully done study in a similar situation measured students' perceptions of crowding in two dormitories with the same floor area and number of students but with different layouts. One had rooms arranged in suites or small groups, while the other had long, double-loaded corridors. In the former, it was found that residents were more familiar with each other, demonstrated ability to cooperate more effectively with others, and did not consider the place as crowded as did residents of the latter.[1]

Such an awareness of how crowding is perceived leads away from reliance on minimum standards. These standards may help prevent some especially severe crowding, but they should not be confused with informed design guidelines.

Territory

Territory can be viewed in two ways. "Primary" territory is the place where possessions are secure and an individual is assured of

---

1. Susan Saegert, ed., *Crowding in Real Environments* (Beverly Hills, CA: Sage Publications, 1976).

continuous control. For many people, this is a whole household. For people in institutions, there is often no primary territory at all, because the institution controls everything, including storage. In group living, usually at least the bed and immediate area around it are primary territory.

Another kind of territory—"secondary"—can be taken over by a group or an individual for periods of time and then left for others to use. Taverns, neighborhood street corners, offices, and phone booths are just a few examples of secondary territory. The shared spaces in most group-living settings act as secondary territory. The kitchen is the cook's domain during meal preparation; the dining room belongs to the women's group while meeting there; the porch swing is temporarily owned by the couple who are alone there.

Making decisions about secondary territories develops negotiation skills. Users must reach agreements about the arrangement and use of shared spaces. The degree of personalization is a good indication of whether or not this negotiation has taken place: if people are willing to care for and protect each other's personal contributions, there is probably some shared agreement.

Recognizing and respecting territory is an important social skill. In group living, if secondary territories are used cooperatively and primary spaces are respected, these conventions can be learned and practiced.

When people cannot control some sort of terri-

2-10. People must have their individual territories to nurture a sense of control and enhance self-image.

tory, they become powerless and marginal. People in institutions have next to nothing to call their own, and their self-image is likely to be weakened accordingly. In a group home each member should have a substantial primary space to reinforce self-worth.

BEGINNINGS—STARTING A GROUP HOME

While the guidelines of this section are slanted toward finding and evaluating existing buildings, they can also be useful in designing a new building. These guidelines provide a basic understanding of the issues involved, but professional advice will often be needed to evaluate the soundness of a particular building, adequacy of electrical and mechanical systems, and to understand zoning and building codes. Users of this section, whether experienced or not, can benefit from other people's experiences, and we have included many firsthand stories. Even so, each person who has been through the process of setting up a group home will say that there was really no way of knowing ahead of time what to expect.

You are wrong if you think that everyone in a group home is incompetent and needs to be treated as dependent children. It may be that you need to have "children" to take care of; you may have unfulfilled longings for affection. But you will not be able to develop an atmosphere of skill development and growth if your true agenda is to have someone dependent on you.

You are wrong if you think that running a group home will solve your money problems. It should not and usually does not. Moreover, if your biggest concern is the income residents bring in, you will fail to see them as lively individuals. The needs of each person will surely not be directly equivalent to the income generated.

You are wrong if you suspect that everyone who has been institutionalized is a rapist, psychopathic killer, arsonist, or child molester and therefore needs to be watched. Similarly, you are wrong if you think that all the residents of your home are just waiting around to be cured, waiting until that day when life becomes "real" again. Some people's problems may never be cured, and it may actually do harm to think in these terms.

The status of group-home residents is special; they are neither blood relatives nor legal owners. They are also not guests. Yet the point of group homes is participation in the real events of everyday life to whatever degree one is capable. The people who operate group homes need special ability and insight to help make this happen.

We suggest that users of this section keep a notebook throughout the search for a house. A tremendously important notebook item is the checklists that you make up yourself. From the beginning, you must confront important issues— what is really important to the group and program for your house? As you read through this section you may find that the most important part of making a checklist is the questions that are raised during the process. Questioning the necessity of a bus stop nearby may, for example, help define the function and importance of transportation for future residents.

Be creative. Look at the examples of questions we have shown in figure 3-1. Some of the questions can be answered with a simple check, while others need a number. Some questions require comparing different places or making choices. We have purposely left the examples incomplete with the hope that this will inspire you to create your own checklists.

In addition to your own checklists, two additional useful items are:

A tape measure: 16 feet is a handy length. Measure room sizes, particularly to determine whether square footage of bedrooms and social spaces meet licensure requirements.

Polaroid camera: Take a few photos of interesting rooms or features and of the exterior; especially useful if you're considering several homes.

Location

A former program director observed that "location of group homes is really an accident of where all the big old houses are in this city." Some areas, always in the most run-down, dangerous parts of town, have become "mental health ghettos." Because it is easier to find and establish homes in this sort of location, the important principles for locating a community home are often ignored and the path of least resistance is taken.

Opponents of community facilities use the concept of saturation either to keep group homes out of their neighborhood or to keep old-line institutions in business. On the other hand, people who

Checklist.   Page 3 of 4.

16. Kitchen. (Cont'd)

Counter: Is there at least 2 feet of useable counter space next to: 1) Refrigerator? __ 2) Range? __ 3) Both sides of sink? __   Total counter length = __ feet.

Adjacent spaces: Is there space within the kitchen or next to it for a breakfast nook? __ Does a kitchen entry mean that people will walk through to get to the rest of the house? __ Is there a full food storage closet near the kitchen?

17. Laundry.

Location. Is the laundry presently located in the basement? If so, is there space and access to plumbing & vent so it might be relocated on first floor? __ On second floor? __

Area. Is there space for:

- A table for sorting & folding clothes.
- Socializing: Making coffee or snacks, watching TV.
- Ironing board, sewing maching, hanging rack.
- Storage: Detergents & bleach.
- Reading, writing, doing homework.

18. Bedrooms.

```
1 2 3 4 5 6
```

1	2	3	4	5	6	
						Windows on 2 walls for light & ventilation?
						Plenty of closet & storage space built in?
						Alcove, nook or "L" shape?
						Walls, ceiling, floor, trim need repairs?
						Walls, ceiling, floor, trim need refinishing?
						Adequate grounded electrical outlets, lights?
						Estimate square footage of each room.

```
1 2 3 4 5 6
```

Special features. Are there special features of the bedrooms which would make them especailly individual & attractive? (These might include a window seat, direct connection into a bathroom, a sloped ceiling, etc.) _____

3-1. Sample design checklists help to identify a group's needs.

otherwise favor the community movement see saturation as responsible for the horrible conditions of some community housing—the welfare hotel, for instance.

Saturation raises qualitative issues that require good judgment. Some apparently poor neighborhoods may actually offer a range of supportive relationships, a network of friendship and mutual help from the other group homes located there. This sort of "saturation," if it must be called that, can be positive and comforting. As that same former program director put it, "It's not the concentration of numbers that makes it bad; it's the spirit that pervades them." Finding a neighborhood with just the right spirit may be the most difficult part of the process of opening a group home.

Safety is one primary concern when considering a neighborhood. If residents will fear leaving the house, there is little point in being in the community. A neighborhood with a reputation for street crime defeats the idea of a community residence. Do not consider a location safe until you have walked it yourself, day and night.

On the other hand, tightly knit neighborhoods and homogeneous suburbs can be just as bad. Many people who have been institutionalized do not conform to the conventions and imagery of such neighborhoods, and to locate them where they have nothing in common with their neighbors will probably result in further alienation rather than integration into community life.

It is far better to look for diversity. Mixed housing types, a varied age group, established residents as well as transients (college students, for instance), a diversity of lifestyles are all positive signs for a potential location.

Access to transportation must also be considered, for it makes possible access to a wide range of necessary support services. And, just as important, it also means freedom and independence. Using public transit is an important urban survival skill, one that is easily lost in institutions.

Group homes are best located within a two-block walk of a major bus line, where buses run regularly into evenings and on weekends. Only rarely will someone just coming out of an institution have a car. Some group homes have a van, and a staff member will drive people around in

groups. While this may be necessary in, say, a rural, farm-based home, unfortunately it perpetuates dependency and isolates residents from the community at large.

The neighborhood services available are vitally important to the success of a group home. If all that people do is sit home and watch television, it makes little difference whether they are in an institution or a community setting. If the group home is within an easy walk of a variety of neighborhood services, the residents will be attracted out of the home into the surrounding neighborhood. Such places should not be expensive, since most residents simply do not have much money to spare. There should be a choice of two or three places within a couple of blocks of the house. Any farther and it is unlikely that they will become a regular attraction. These places must allow people to linger. Fast-food restaurants are definitely out, but such places as laundromats or variety stores, a local park, car wash or community center are attractive. It is especially nice for residents to be able to pick up extra coffee or cigarettes on the spur of the moment. Keep other services in mind too. Medical and dental clinics, a bank, church, and a post office should be readily accessible.

What opportunities are there in the neighborhood for employment? Are there places to meet with other people and to be valued by neighbors? The answers to such questions will help determine the suitability of a neighborhood.

House Types

Most people start with the program and try to find the house to suit it, but others start with a house and let it help shape the program. If you start with the house, the size, of course, largely determines the number of people in the program, because of the size and number of bedrooms and shared spaces. A house with a large kitchen and laundry area is perfect for a program aimed at learning living skills. A small kitchen would do for a program for skilled clients who rotate kitchen chores without much supervision. Spend time imagining the sorts of program activities the house would best support; let it suggest possibilities, by its size, layout, and location.

If you allow the program to determine the house, whatever your program model—vocational, psychosocial, supportive peer group—you must understand the need for both private as well as shared space.

Imagine the round of daily events. For instance, if all residents leave at the same time for a sheltered workshop, the entry hall will be heavily used at given times. Also everyone will be wanting to shower at the same time so you will need more than average bath facilities. You also may need spaces for staff, live-in or not. If you do not provide for this ahead of time, there may be a constant battle for the choicest space later. Use your program activities to shape the size and type of spaces you need.

3-2. Large old houses provide ample room for group and individual activities.

Large Old Homes

Big old homes provide places for group members to be together as well as for them to be apart— critical for good group functioning. Construction costs almost preclude duplicating the variety of gracious social spaces such houses offer—dens, breakfast nooks, libraries, and sun porches. Circulation is usually in hallways, rather than through rooms; the privacy this affords is especially important since these large homes may easily accommodate fifteen residents.

These old houses do have drawbacks that must be considered. Many rooms are simply too large to furnish economically, yet they are difficult to remodel without destroying the house's integrity. The result is often a bare living room that resembles a hospital dayhall. Old homes are also notoriously energy inefficient. They can be expensive to operate, particularly if the house has gone through a long period of neglect.

People running lodge programs see real advantages to large old houses. A big house can anchor a program, and give members a sense of place. In the Fairweather lodge model, residents form a sustaining and self-governing group, taking responsibility for running the lodge, operating a business, and handling finances.[1] The role of program staff is deemphasized. Lodge programs generally emphasize the supportive peer group

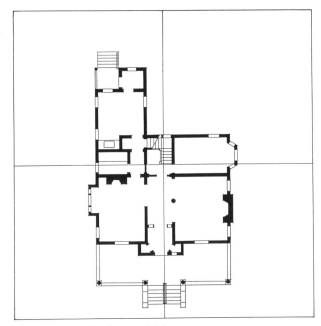

3-3. Layout of a large old house.

combined with viable nonpatient community roles such as employee, tenant, worker, consumer. Operating its own business, a lodge gives people

1. George W. Fairweather, ed., ''The Fairweather Lodge: A Twenty-five Year Retrospective,'' *New Directions for Mental Health Services*, no. 7, 1980.

income, a place to live, and the responsibility of employment. Residents run the lodge themselves, cooking, cleaning, and making decisions and rules.

Smaller Houses in Cities and Suburbs

In smaller houses, circulation is through the rooms—the newer open-plan house (usually found in the suburbs) and older city houses. While suburban houses have the drawback of being located in car-oriented neighborhoods that lack many important services, they generally have more up-to-date equipment and are more energy efficient. Compared to large old homes, these houses have fewer different spaces. Circulation is through rooms connected directly one to the other, with less privacy as the result. Such a plan tends to limit options for a larger group but will work well with long-term programs for smaller groups operating in the manner of an extended family.

There are other options for group living. Perhaps two small houses in the same neighborhood would work better than one large one. Consider physically connecting two small adjacent houses by building a link. A new bedroom wing can be added to an existing small house.

Apartment Buildings

Apartment buildings, particularly in older neighborhoods, make attractive group homes. They are zoned for unrelated people and are located where there is already a concentration of diverse tenants who may move occasionally—unlike the rigid single-family area. One agency director started out by looking for:

. . . two group homes and an apartment building. We looked and looked and looked for two years or so and never found anything we could consider suitable for a group home. Everything we found was either too small, too large, too expensive, or not in the proper zoning. . . . We had a commitment for the grant money [and] decided to put it all into apartment buildings because they are relatively easy to find. And we found two which were perfect, which is where we are now.

As we found our reality changing (that good houses are impossible to find), we found our philosophy changing also. I know that's backwards but that's the way it happened. We decided, clients are going to appreciate this much more: the privacy, the fact that

3-4. Small modern houses are usually more energy efficient than older homes.

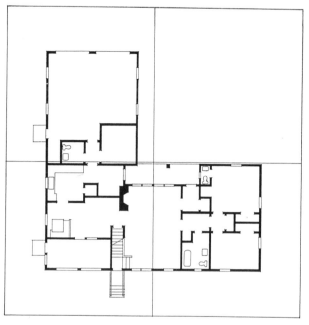

3-5. Layout of an open-plan modern house.

3-6. Apartments offer privacy and the opportunity to master daily tasks.

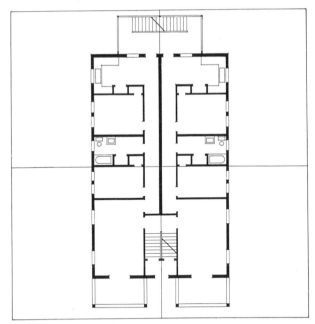

3-7. Layout of an apartment building.

they are in smaller groups, they don't have to accommodate and adapt to so many people. They will probably be closer with each other, give each other more support, and feel a sense of ownership much more in a smaller unit. And that's the way it worked out. The clients tell us it's far superior to any group home

they've been in . . . It's because there isn't all that constant traffic, both physical and psychological.

Most apartments offer a lot of privacy. For programs emphasizing peer group support, one apartment is sometimes set aside for group activities or for use by staff.

It is often assumed that residents will not be able to cope with day-to-day tasks such as cooking. However, some apartment operators say that there is no better learning situation, with the appropriate level of support.

Duplexes

The popularity of two-family houses has waned over the past generation, and such houses are more available, and, often, less expensive than single-family houses. What makes them attractive as community homes? First, some zoning laws consider up to four unrelated people as a family (although this varies from place to place). This means that a duplex may typically accommodate up to eight residents without any special zoning variances. Secondly, the average room size in a duplex will be smaller than in, say, large old mansions. Smaller bedrooms mean more individual, private spaces and fewer shared bedrooms. A third advantage, easy to overlook, is that a duplex will usually be fairly compact and therefore more energy efficient.

Codes and Regulations
Zoning

The local government, not the state, sets up restrictions on how land may be used. In an area zoned for single-family houses, for instance, apartments are not allowed. Local governments may set up zoning procedures as they choose, so the rules and procedures that apply in one area may be different from another. If there is no local zoning department, the best place to find out about zoning is probably the local building

3-8. Duplexes are sometimes more available than single-family homes.

3-9. A zoning map indicates how buildings in particular areas may be used.

department. Most municipalities have a zoning map available, showing each piece of property and its associated zoning (along with allowable heights, square footages, and other requirements).

Complications with zoning center on the fact that it is often used as a mechanism to keep group homes out of particular neighborhoods. A zoning hearing may be the only public forum for discussion of a prospective group home, and the atmosphere may well include hostility and bigotry. Generally, neighbors fear increased crime, and property owners fear decreasing values, despite many studies that have shown that neither happens. Complications begin with the inconsistent terms used in different places. Even the definition of *group home* varies from one place to another. Some municipalities will consider residents members of a single family and allow a group home in an area zoned for single-family dwellings. More frequently, however, the opposite is true. Even apartment buildings may not be acceptable if, for instance, meals are to be served in a central place. In some areas the building would then be considered a boarding house and be prohibited in an area zoned for apartments.

Fire Codes

Fire codes cover such things as the construction materials used in the building, the arrangement of exits (particularly from bedrooms), fire protection equipment, and the storage of hazardous materials.

Code compliance is assured through inspections. In making judgments, different authorities will use their own inspection processes, often employing a combination of state or national codes. In addition to fire safety, inspectors may also be concerned about accessibility for disabled people or about habitability—plaster and paint, serviceable floors, condition of the exterior of the building, and so on.

A group-home sponsor who is familiar with the codes stands a much better chance of keeping the house intact and saving money on required alterations. This is because the codes often give two or more options to accomplish the safety measures intended. The precise requirements vary with the locality and the individual inspector's interpretations. For example, some codes will permit more lenient exit requirements for a masonry house than for a wooden one.

The three usual sources of community fire regulations are the state fire marshal, the local fire department, and the National Fire Protection Association. Decisions about which codes apply typically come from three different bodies: the local building department, the state group home licensing agency, and often, from the funding source (such as Medicaid).

Of the three sources of fire regulations, the most basic is the code published by the National Fire Protection Association (NFPA), known as NFPA 101. It has been made quite accessible through an illustrated handbook: *Life Safety Code Handbook*. (This and many other fire safety publications are available from the National Fire Protection Association, Inc., 470 Atlantic Avenue, Boston, MA 02210.)

The first step in code compliance is deciding on the correct "occupancy" designation. Group homes will usually be classified under one of three occupancy designations: "Health Care Occupancy," "Board and Care Homes," or "Lodging or Rooming Houses." (Be aware that these NFPA designations are *not* the same as those used in other codes, such as state fire codes.) Each of these occupancies is covered in a full chapter of requirements in *NFPA 101*. In addition, "Health Care Occupancies" and "Lodging or Rooming Houses" are divided into additional chapters for "New" occupancies and "Existing" occupancies. The best way of deciding on the correct occupancy—and, in turn, knowing what requirements must be met—is to read the handbook.

If the issue of occupancy is determined correctly, fire codes ought to be welcome—they help save lives. "Board and Care Homes" may be the correct designation for most group homes, as the fire safety measures required are tailored to the specific self-preservation abilities of residents. Unfortunately, this is a new chapter in *NFPA 101* and may not yet be understood by some fire inspectors.

Too often it is necessary to deal with inspectors who, once they know that the house will be for a vulnerable group, will jump to the conclusion that the occupancy must be "Health Care." If this decision is allowed to stand, some very institutional (and very expensive) measures must be taken to meet the code.

One very common institutional measure is the installation of a commercial-type sprinkler system of the type prescribed in *NFPA 13*, "Standards for the Installation of Sprinkler Systems." Usually this requires very large, visible supply lines and "heads" on the ceiling of every room and closet. Sprinkler heads are activated by heat and, in some systems, also by smoke. They are very expensive because of the elaborate piping, the difficulty of installation, expensive fixtures, costly maintenance, and on occasion, special pumps. If they are accidentally set off, they may cause serious water damage: this fact leads in turn to additional house rules to avoid setting them off accidentally—giving the home an additional measure of institutionalization.

Under the "Board and Care Homes" occupancy chapter of *NFPA 101*, a far less expensive and less intrusive sprinkler system is allowed. It is for one- and two-family dwellings and is prescribed by *NFPA 13D*, "Sprinkler Systems— One- and Two-Family Dwellings."

While sprinkler systems do save buildings, there is some debate about their effectiveness in saving lives. There is no such debate, however, about smoke detectors. Properly designed and combined with safe egress routes, they do save lives. Smoke detectors were designed in response to the realization that smoke, not heat, kills most victims of fire. They are inexpensive. Every operator should welcome the security of a good smoke detection system whether required by law or not.

Another common institutional measure is the complete enclosure of stairs leading from the first to the second floor. This measure is usually required because every floor (including basements) must have two exits to provide an alternate escape if one route is blocked by smoke or heat. The critical issue is the nature of the exits. There are situations, depending on occupancy designation, in which only one of the two exits from a floor needs to be enclosed. Porch roofs and ladders are also allowed in some egress plans. A fire escape itself might be designed as a balcony, actually enhancing the house. If there are bedrooms allowed on higher floors at all, they most certainly must have two separate escape routes.

Since most fire deaths are caused by smoke, it is a common requirement in institutional occupancies that surface materials be fire resistant and that they produce little smoke when they burn. In those homes where this is not a requirement, it is still worth paying attention to, even if this means excluding some materials from the house. In addition to wall surfaces, furnishings such as carpet, upholstery, and drapes should be carefully selected for fire resistance. Many of these surface materials have been given a fire rating—Class A, B, or C. Class A materials contribute the least smoke and fuel to a fire and are, therefore, most desirable, especially in those areas and furnishings where fires are likely to start—beds, closets, wastebaskets.

Many popular books on home decorating describe the fire-related characteristics of fabrics. Fabric or upholstery stores may also have this information. Fabric dealers can send fabrics to a finisher who will chemically treat them and provide certification. These treatments are not permanent and may have to be reapplied after washing or dry-cleaning. Some fabrics are inherently fire-resistant—wool, felt, certain synthetics—and do not require treatment. A conscientious operator will keep records of materials purchased and treatments given, whether required to do so or not.

Finally, the best means of preventing fire and death lies within the residents' own concern for self-preservation. Having a voice in making rules—everyone agrees that there should be no smoking in bedrooms—and having their enforcement be a responsibility shared by the group means that the rules will be taken to heart.

Other Codes

Health codes may focus on food preparation, sometimes requiring institutional equipment such as three-compartment kitchen sinks. Sealed, sanitary surfaces in kitchens, as are often required by health department regulations, may help control disease. This sometimes means that in looking for a home, you may ignore the present condition of the kitchen, since it will have to be replaced anyway. In one home where the kitchen had to be virtually rebuilt, the operator was able to locate a new kitchen in what had been the dining room. The old living room has become a

nice, large dining room, and the old, small kitchen is now used as a combination breakfast nook and mud room.

Building codes are concerned with the building structure, wiring, and plumbing. Building inspectors are sometimes the bearers of unexpected demands. While excessive cheeriness in the face of adversity can wear thin, consider this: even if drastic alterations are required before moving in, there may be some opportunity to go beyond merely complying with a code. For instance, if rewiring is required (which must be done by a licensed professional), the job can be broken into parts. Residents can be involved in cutting channels in plaster walls to accept new wiring. (This is usually cheaper than either snaking new wiring behind existing walls or mounting new wiring on the surface, even if you have to pay a professional to patch the channels.) Another solution is to design a ledge system to be mounted along walls which can carry the wiring behind. This will also protect walls from chair backs and double as a handy ledge for displaying personal items.

Other jobs can be broken into parts, using professional time sparingly. Drywall taping requires professional skills; hanging drywall does not. Overall home insulation to conserve energy requires professional advice, but it is quite reasonable for amateurs carry out some of the installation work (and it is also reasonable that they be paid for their work). Weather-stripping doors and windows or installing insulation on the inside of basement walls can also be done by amateurs.

There are impossible situations. One home was required to install a copper drainpipe from a plastic laundry tray to an open floor drain. PVC plastic pipe would have been perfectly adequate and would have cost about one-twentieth as much. Such things happen.

Active Places

One persistent problem in group homes is the use of shared spaces. Too often they are limited to one activity. Some of this can simply be avoided at the beginning by careful selection of a house.

Whether through character, layout, floor plan, or design, some houses are more livable, usable, and comfortable than others. A few basic characteristics affect activity—natural lighting, good ventilation, easy traffic patterns, and room sizes that are neither too large nor too small. Keep these in mind when viewing any house.

Having windows on two sides of a room generally ensures good air circulation and also means natural lighting will be even and without glare.

By virtue of the layout of rooms, it is easy in some houses for residents to be in touch with the activities throughout the house; other houses can be confusing and cause residents to feel isolated. The ease and comfort of knowing where activity is or of being able to find a quiet corner is important where people live as a group.

When a room must be used as a traffic lane—to reach another room beyond—options for use and privacy are more limited. Small rooms may be limited in potential while large rooms can be awkward.

Good space can be used for different activities by different-size groups during the course of the day. You do not need a separate room for each and every activity or group. A room with two parts (L-shaped or with an alcove or niche) offers easy flexibility.

To work well for a larger group, a house must have several activity centers, separated from each other. Without some choice, people will feel always on top of one another.

The television should not be such a center, for it promotes passive activity. Nor should the place where the whole group gathers for a meeting be considered an activity center. Generally, finding a space for all to meet is not a problem. More often, the problem is that large activity centers seem empty and "dead"; keep them small and lively.

Visualize several (three to five) centers: the main entry with people arriving and leaving; the kitchen; the hearth, where everyone congregates after dinner; a place for noisy games and music; a place for messy hobbies—repotting plants or painting a chair.

When walking through a potential house, room by room, try to envision these centers. Do not think only of rooms with furniture but of the activities each center will support.

The Kitchen

It makes no difference whether the kitchen will be run by a professional cook or by residents of the house; it can automatically become a desirable center of activity. The problem is finding ways for people to be involved without creating new conflicts—getting in each other's way or stepping on the cook's territorial toes. This is an interpersonal problem that cannot be solved with locks or closed doors. Such institutional thinking perpetuates feelings of exclusion. Instead, areas within the kitchen can be defined so that several people can be doing different things at the same time. For instance, in one area two or three people can work together, scraping plates, washing dishes, removing garbage, and so on, while in another area, out of the way, others sit at a table or booth and have coffee. In a small kitchen, a pantry or adjacent space can be opened onto the kitchen and used as a snack area.

Dining Rooms

The room that works best for dining may not be the dining room. In too many cases, the dining room resembles a cafeteria lunchroom, cramped with an extra table or two. Worse yet, between meals the room is liable to stay unused.

3-10. Kitchens can be multifunctional gathering places.

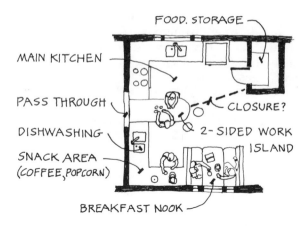

FOOD. STORAGE

MAIN KITCHEN

PASS THROUGH

CLOSURE?

DISHWASHING

2-SIDED WORK ISLAND

SNACK AREA (COFFEE, POPCORN)

BREAKFAST NOOK

3-11. Defining areas in a kitchen can enable more people to use it at one time.

3-12. Home workshops allow residents to be involved in house maintenance.

In evaluating a house, consider using a different room for meals, say the living room. There residents may not be quite so crowded, and the room lends itself to other uses at other times. It can be furnished with several small tables instead of one large one, and a piano and other furnishings will extend its use throughout the day. The original dining room can then be made into a more cozy, intimate living space in which chairs and other furniture can easily be arranged for comfortable conversation.

Such an approach is true for every room in a potential group home. Do not feel bound by what the room was originally used for, but rather allow spaces to fit the needs of your group.

Workshop and Storage Space

Having a shop nearby means residents can easily be involved in regular maintenance of the house, especially those little day-to-day things that need attention—mounting a window shade, replacing a faucet washer, making a shelf for the kitchen. Giving people such responsibility not only increases their self-confidence and competence but also keeps the place from running down. The space chosen for the workshop—whether in the basement or garage—should be dry and warm in winter and have good ventilation.

Appropriate storage space enhances house

activities. Storing bulk foods in the basement releases the pantry for other uses, for instance. Do not plan to use space for dead storage that could be used by people.

Entry

Try to imagine from where people will really be coming and going. It is often not through the original front door, but wherever it is, it is an especially important place to group-home residents. They are liable to have to wait there for a ride or a visitor, often there is no place to sit or to see the drive. There should also be plenty of space for coats and boots.

Creating a mud room with a view would be a successful solution. A breakfast nook with a view would also be very effective. There, one might have a cup of coffee and say good-mornings while waiting for the downtown bus.

Wherever the entryway is, it is an active place, for keeping watch over comings and goings.

Laundry Areas

Do not plan to have your washer and dryer in the basement. The laundry room is an activity

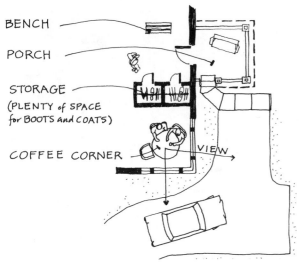

3-13. An entryway that is visible from several locations is more convenient for residents.

center that is much too important to be relegated to out-of-the-way places. Try to locate laundry equipment on the main floor in an area that is visible and large enough to promote washing as an opportunity for mutual help and learning. It becomes a convenient place for several people to be together and work.

Allow adequate space for folding clothes and storing laundry supplies. Furnish this place carefully with a table and a few chairs, because this setting is a good opportunity for lively conversations and other activities such as card playing or letter writing while waiting for clothes to dry.

You will need a water supply and drains and appropriate gas and electrical service, as well as a proper vent. These can be installed economically and should not restrict your choice of locations.

If you do have decent basement space, consider locating a television set or perhaps a pool table there. Either one is a natural magnet for activity, and neither needs to be centrally located.

Private Spaces

Of all the factors that might concern you in the environment of a group home, there may be none more important than privacy. It is the feature that most clearly distinguishes community life from institutional life.

In the very simplest terms, privacy means having control over contact with other people. Territory, a closely related concept, means having control of a place. Buildings give privacy by marking territory. Most of the elements of an everyday house lead a double life. Each part serves some obvious function and, at the same time, helps to mark territory. Outside the house, a slope down to the sidewalk helps drain water away from the house and also marks the edge of the front yard, defining the boundary between public and private property. The porch provides shelter from the rain but not for strangers unless invited. The front door may keep out the cold, but it is also a marker of territory under the control of the people inside. Such territorial markers are so thoroughly built into the environment that their double meanings are easily overlooked.

Of course, these marks do not have inherent meanings of their own; they form a language that must be understood, shared, and respected. A door is not effective as a territorial marker if anyone can enter without knocking. People become angry if these marks are violated. In group homes, the rules that govern staff should also govern residents. Nobody, not even owners or operators, should be allowed to intrude in private spaces.

Keeping people apart is just one aspect of territorial markers. Because barriers and boundaries are a means of control, people use them to live more closely together and interact more easily. Talking across a table or over a back fence is often more comfortable than sitting or standing out in the open, exposed.

People who have a private place are more likely to interact frequently with others than are people without privacy. In institutional life, misguided ''control'' types often push patients into the dayhall to ''interact.'' This does not work, of course. Instead, by creating a situation of no choice, they actually cause people to withdraw emotionally.

Clearly defined territories also reduce incidents of fighting, verbal abuse, or just repressed anger. Without definition, territory will be encroached upon by staff as well as residents.

Interpersonal conflicts and withdrawal are not

corridor circulation will give more privacy than a compact house where circulation is through the rooms. Three or four activity centers give more choice than only one large, shared space. Within individual rooms, look for alcoves, nooks, and niches. Look also for out-of-the-way places—usable stair landings or basements. Each of these provides additional opportunities for private places. The need for ample storage space may not immediately seem to be a privacy issue, but having appropriate storage areas means that more attractive, accessible spaces will not be wasted.

The privacy available in any shared space is also an important consideration. As one expert notes:

The spatial interrelationship of the public rooms is also of interest. Spaces that are visually accessible are preferable to discreet, cellular spaces. A resident who is tentative in his capacity to relate to others may be comfortable reading alone in a corner of a room. If that room is visually open to connecting rooms—as through a large doorway or archway—then the resident will not be totally removed from others. This represents a psychosocial principle expressed through architectural design: visually interconnected spaces

3-14. The two women on this porch enjoy the privacy it provides.

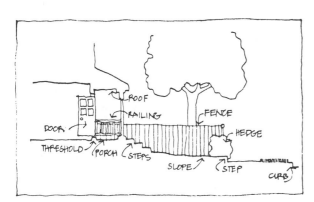

3-15. This layout enables residents to have privacy from outsiders without remaining indoors.

necessarily caused by psychological problems, since either may be a result of badly designed or badly used space. To avoid blaming the victims, it is imperative to distinguish between the problems of people and of places.

In selecting an appropriate house for a group home, look for fixed features in the layout of a house that give privacy. A house with wings or

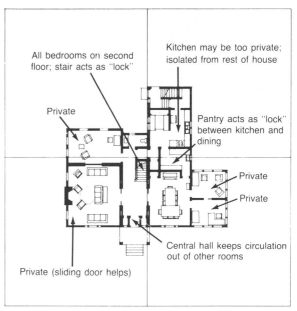

3-16. This layout offers many private spaces that can be used for shared activity as well.

create a capacity for privacy and separateness without isolation."[2]

Houses in which the only private space available to residents is the bathroom are inappropriate. By the same token, a staff office ought not to be the only place in which staff members can find privacy.

Look also for buffers between spaces. Vestibules, entryways, and corridors can separate one room from another. There are other ways to control access and sounds—such as solid, tightly fitting doors. However, dutch doors, glass and sliding doors, indoor windows, and pass-throughs will not control sound adequately. They are valuable for other reasons, for they do offer a realm of important choices.

Having territory allows for a personal style, one that may not always conform to others' ideas of how things should be. A commitment to each person's right to privacy also means accepting—even celebrating—individual differences. Obviously, filth and roaches are to be discouraged, but there must still be room for a personal style to bloom.

Privacy is essential for normal behavior. It is not to be withheld or given only as a reward for good behavior. After all, it makes no sense to withhold necessary supports and then expect normal behavior. Privacy and territory give choices, power, and control, and these elements are basic to human growth and development.

Ownership

People enjoy a sense of ownership when they are part of the decision-making process. Starting a group home is more than just buying a building; it also means starting a group. The house itself can be a means of expression for the group. Through it the group can feel ownership and control.

Establish an open style from the start by organizing a client group before the home is opened or even found. If this is impossible, you can ask

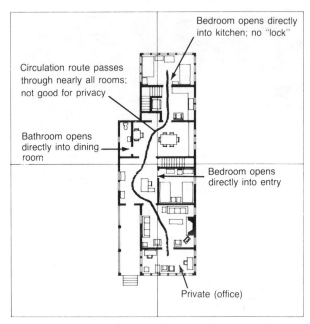

Bedroom opens directly into kitchen; no "lock"

Circulation route passes through nearly all rooms; not good for privacy

Bathroom opens directly into dining room

Bedroom opens directly into entry

Private (office)

3-17. This layout does not offer as much opportunity for privacy as that in figure 3-16.

for help from two or three prospective residents. Together, visit houses you are considering and other group homes in your area, and walk the neighborhood. Do all this together, sharing reactions.

If choices must be made about renovations, involve residents in both the decision making and the actual work where feasible—painting, cleaning, and getting the house ready for occupancy. In the bedrooms, residents should be allowed to choose the furnishings from those you have available.

Also involve the group in planning the program, making rules, and deciding on day-to-day responsibilities. Let residents have a voice in decisions about space use and chore schedules.

The Issue of Security

A robbery or break-in not only destroys the sanctity of a home but threatens the personal security of individual residents. Nor can a resident feel comfortable in a house where sexual exploitation or petty theft goes unchallenged. On

2. Richard D. Budson, *The Psychiatric Halfway House* (Pittsburgh: University of Pittsburgh Press, 1978), p. 159.

the other hand, excessive security measures (complicated alarm systems, glaring mercury vapor lights, identification badges, closed circuit cameras) are all so obtrusive that they too can rob the home of its sanctity. It is essential to achieve a balance between security and a feeling of openness.

The first defense is the vigilance of the group. Work together to survey problems. Involve the residents in deciding on security measures and their enforcement. Nothing equals dedicated, on-site vigilance, and who has a better reason to be vigilant than those who feel ownership? Cultivate feelings of ownership.

Some houses have better characteristics for security than others. A bedroom on the first floor, near the entry, for example, is a plus. Outside doors and basement and first-floor windows are all vulnerable, and should be where they can be seen by occupants as well as by neighbors and passersby. A house with single bedrooms is better for personal security than is one that offers only multiple-occupancy bedrooms. But all in all, the characteristics of the house will not—by themselves—ensure a secure home.

No matter what measures are taken, there may still be some problems. For one thing, it is easy to overlook (and, therefore, fail to protect) some valuable object. Two hundred pounds of beef in the freezer may be very attractive to a would-be thief. The issue is not whether there will ever be a problem but how to deal with problems effectively while, at the same time, maintaining the strength of ownership processes.

Skills and Responsibilities

Many homes, both owner operated and agency run, fail to take advantage of the opportunities of daily life for building skills. In these homes, all cooking and cleaning—even residents' personal laundry—is done by staff. Residents are driven everywhere in the van and have their appointments made for them. As a result, residents mostly sit around all day in front of the television, perhaps smoking but with nothing else to do. This is far from the original spirit of the community movement.

3-18. Bulletin boards and datebooks near phones enable residents to take messages and meet appointments, important living skills.

Another result of this situation is that some home operators become bitter about a perceived lack of gratitude on the part of residents. "I do everything for them," one operator confided, "and they just don't seem to appreciate me." "Doing everything for them" does not really benefit anyone.

Instead, given opportunity and appropriate support, people can assume responsibility for their own lives, decisions, and choices. This approach recognizes that all people, even institutionalized people, are capable in different ways and to different degrees. Said one resident, "Use experience as a teacher, don't practice how to live. Get on a bus and learn to ride one. You just can't teach that stuff in an institution." And an operator discovered ". . . that we could count on people to take care of themselves more than we thought they could. In all areas—shopping, cooking—they needed only a certain amount of supervision and assistance and not taking over all those responsibilities. The basic thing is we were underestimating clients all the way around.''

Structural Soundness

This section should not be a substitute for expert advice, but it will give you a basic understanding of what to look for when considering structural soundness. It may at least help you eliminate overly run-down houses. If a particular house passes these tests and fulfills the requirements of your checklist, you should then consider consulting a professional before purchase.

Inspect the exterior of the house first. Step back and look at the roof. Does the roof line at the peak (the ridge) appear straight and true, or does it sag? A sag is a definite warning of seri-ous problems: the foundation or the structure itself may be failing. Looking out a second-floor window to a porch roof can tell you a lot about the roof as a whole. Are the shingles worn smooth with age? Are they cracked? Are shingles missing or beginning to curl up at the edges? This is a sure sign that the roof is old and will need replacing.

The biggest problem with gutters and down-spouts occurs when they are not properly main-tained and get blocked with leaves. Check the ground around the perimeter of the house for signs that water is running directly off the roof; this often leads to moisture and structural prob-lems in the basement.

Gutters and downspouts feed into storm drains. If the drains are blocked, water backs up. Excavation may be required to correct this.

Check brick or other masonry structures for

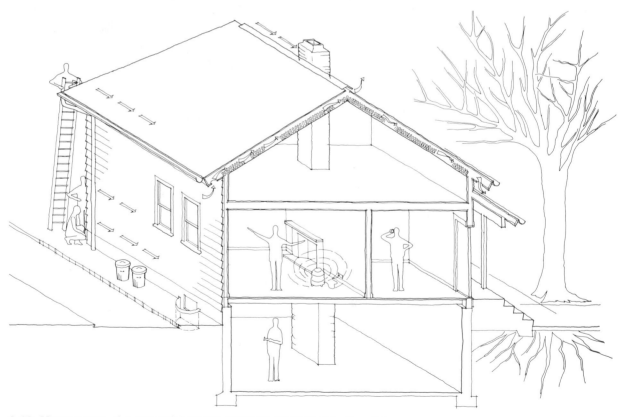

3-19. Many areas of a potential house must be inspected before it is purchased.

mortar missing from joints and for spalling of the surfaces. Both problems allow water to penetrate the wall, where, in cold weather, it will freeze and expand. This expansion forcibly breaks out more of the wall and eventually destroys it. Mortar can be replaced through tuck pointing (which is expensive and, if needed, should be planned for in the budget). Spalling means the surface of the brick itself is crumbling away. If you find considerable spalling, you ought to look for another house.

Moisture is the enemy of wood siding, coming either from the outside through cracks or as condensation from the inside. (Water condenses within the wall when warm inside air meets the cold outside structure.) Look for signs of peeling paint on walls outside of kitchens and bathrooms; these two rooms are loaded with moisture. Rotten and warped boards are other problems that may be caused by moisture. An icepick is handy for probing suspect boards to see if they are mushy and rotted. Soft spots may also indicate termite damage.

Sight along the walls for bulges. If the walls are not straight and true to the eye, get a professional opinion on their stability.

Inspect porches to see if they are rotted, if

3-21. Missing mortar in brick and masonry may allow water to penetrate the wall.

3-20. Gutters and downspouts should be checked for proper drainage.

3-22. Peeling paint may indicate a wall has been exposed to too much moisture.

they sag, if they are missing floorboards, or if the edges of the floor are worn. Look to see if the ceiling boards are collapsing or sagging around the tops of posts or columns. A seemingly small problem like this can be expensive to repair. Do not forget to consider accessibility—a disabled person should be able to get into the house.

Obvious water stains or mold indicate a wet basement. Stains could mean that storm drains need cleaning, or they could portend serious foundation problems. Painting or applying some sort of sealer usually will not solve the problem; it will just cover it up.

Sight along a foundation wall. An obvious inward bulge at ground level means the wall is not strong enough to hold back the pressure of the earth pushing from the outside. This problem will be expensive to repair.

Check the interior of the house as well. Bring along a flashlight to inspect dark areas, such as basements. Horizontal cracks in the basement walls generally indicate that water is building up outside the wall. Correction may require grading the ground surface away from the house and repairing downspouts and drains. Vertical cracks in the basement wall mean one corner of the foundation has settled more than the other. If settling has not stopped, you can be sure the problem will get worse.

Added floor jacks or supports indicate that there has been a problem in the past. Do you think it has been corrected? Have someone go to the floor above and jump a few inches. Does the floor flex badly? If so, find another house.

Use your ice pick to probe band joists and sill plates for signs of rot or termites. Soft wood means trouble. Band joists and sill plates are the structural wood members that are in contact with the foundation wall.

A sagging door is usually a sign that the foundation has settled. The real question, though, is whether the sagging is still in progress. If the door sags, you will also see cracks in the plaster or the door frame molding. Are the cracks old or new? You can tell by looking at the paint—has it cracked since the last paint job? If it has, then the settling is still active, and the house has real problems that are expensive to repair.

Cracked plaster in the walls of rooms above is a good sign of settling in the foundation below. A crack that was patched or painted over and has not reopened may mean settling has stopped, a good sign.

Open and close all the interior doors. Are they square with their frames? Do floors fit tight to baseboards as originally built?

Check the floor. Be sure to check for slopes or bulges. Are there holes, loose boards, splinters, or sloppy patches? The finish must also be intact, particularly in kitchens and bathrooms, where sanitation is a real concern. Check for cracks and excessive wear and tear.

Inspect the ceiling. Look for plaster patches, cracks, and water stains. Dropped ceilings may be used to hide problems, so be sure to look above them. Sponged, textured plaster is also often a coverup. Keep in mind that a proper repair job is a sign of good maintenance to be valued; a poor repair job is a sign of continuing problems.

Try out the plumbing. Turn on several faucets at once to ensure there is a good blast of water from each. Fill up every sink to make sure they drain; flush all the toilets and run all the showers for the same reason. Check to see how long it takes for the water to get hot. Also inspect the plumbing pipes. Galvanized pipe is probably older and will need to be replaced. Over the years mineral deposits tend to form in these pipes, restricting water flow. Copper pipe in older structures is preferable to galvanized, as mineral deposits do not build up in copper pipe.

Carefully examine the house's electrical system. If the house has grounded outlets, a new circuit box, and wiring in conduit, the system is acceptable. If you see knob-and-tube wiring, a fuse box, and two-prong outlets, get a professional opinion on the cost of upgrading the electrical system before proceeding further.

You need to know if the wiring is up to date. Be aware that old wiring (knob and tube) is not necessarily unsafe, but that if the wiring is old, it must be checked carefully. We all use more appliances nowadays, and old wiring systems cannot adequately handle today's larger power loads. Check for splices in wiring. These indicate wires have broken and have then been patched together. In addition to indicating that all of the wiring may be older than is safe, splicing can be dangerous in itself in that fires may result from overloading this weakened point in the wiring. New wiring is usually installed in metal conduit for safety. If you see wiring not in conduit, be sure to check with the local building authority for requirements.

Up-to-date service panels have at least ten circuit breakers; more are required for a very large house. It is not uncommon for a new house to have 20 circuits. Old fuse boxes with only six circuits cannot handle today's electrical loads. Keep an eye open for extension cords, double plugs, and other signs that the current user is finding the electrical system inadequate. Check to see if outlets are grounded. An ungrounded outlet has only two slots for each plug; a grounded outlet has three slots. You can be pretty sure an ungrounded system will need updating. GFI outlets should be installed in all bathrooms and are sometimes required near kitchen sinks as well. GFI means *ground fault interrupt*. These outlets will interrupt the flow of electricity if there is a faulty ground, preventing an electrocution. A GFI outlet will instantly shut off power if an appliance (hair dryer, electric shaver) falls into the sink or tub.

Examine the furnace to see whether it will soon have to be replaced. Open up access doors and peer in with your flashlight, checking for rust or other signs of neglect.

Check the manufacturer's label for serial and model number. A simple phone call to the dealer will usually get you the age of the furnace. Twenty years is about as long as you can expect any major piece of equipment to last.

Do not forget that water heater capacity is important—you may need to replace a small, old heater. The twenty-year life span applies to water heaters too. You will need a water heater with a high recovery rate (usually listed on the tank)—an indication of the time it takes to heat the whole tank of water.

Look carefully and think about what you see during your inspection tour. Problems with the physical structure should be visible to the naked eye—walls not vertical, floors not level, obvious settling. Poor maintenance is also obvious—broken stair treads, missing window panes, wobbly handrails, irregular sidewalks. Look for signs of rodents or pests, for trash and refuse. Common

sense might steer you away from somebody else's irresponsibility.

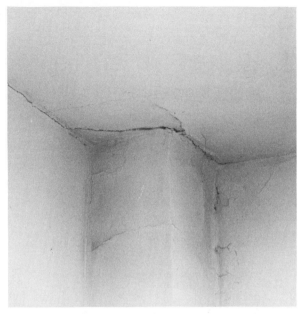

3-23. Cracked plaster near ceilings may indicate that the foundation is settling.

3-24. Porch floors with worn edges may be difficult to repair.

Equipment and Energy

Not surprisingly, a big proportion of operating budgets must go to energy costs. Heat and light, appliances, hot water, and ventilation are all costly. With careful study, you can learn about a furnace, electrical wiring panel, hot-water heater, or plumbing. On your own, you can investigate insulation, storm windows, and weather stripping. Budget billing is available from utility companies. Most local utility companies now offer a free energy audit or otherwise provide consumer advice. But before you really commit yourself to a house, if you have any doubts, you may want to have the advice of someone who has more experience and knowledge.

Find out if the house is insulated. To look into the walls for insulation, unscrew electrical outlet covers (after making sure that electricity is off) or pull off a baseboard. You will see the insulation if it is there; if it is not, plan to add it. If you find insulation, use a ruler to check if its depth is adequate.

It is hard to tell if a finished attic has been insulated, and if it has, if the insulation has been correctly installed. Ventilation must exist between the insulation and the outside structure. In a properly done insulation job, you will see air vents under the roof overhang in combination with either gable vents, roof vents, or a ridge vent at the peak of the roof. Without ventilation, water vapor from inside the house will condense on the cold wall or roof structure and drip into the house.

If an operating hot water heater is too hot to touch comfortably, it is not well insulated. Newer models are generally better insulated.

Unwanted air coming into the house from outside is called *infiltration*. Infiltration negates the value of insulation. Warm air can also leak from inside the house. Check for gaps or holes, particularly in the basement. Look at the sill plate, where the foundation meets the structure above. Attics are notorious for allowing leakage—especially around chimney collars. Loose, badly fitting windows and doors let a lot of air and heat pass through. Tightening and weather-stripping doors and windows are perhaps the most cost-effective measures you can take to save energy. Make sure all glass is tight, with the putty in good condition. Also check the outside of the window frame for caulking where the frame and wall meet.

The house should have storm windows for winter. Even if it does, be careful—aluminum frames let out a lot of heat. Old-fashioned wood or newer vinyl-covered frames are preferable to aluminum.

You should also inspect the house for adequate ventilation. A house that is too weather-tight will have moisture problems and will be uncomfortably stuffy. Fresh air is imperative, particularly where there are cigarette smokers. Check to see that windows operate properly, that they are not painted shut and that sash weights and cords are intact. Baths and kitchens should be vented to remove excess moisture. To increase ventilation without losing heat, you can install an electronic air filter or a heat-exchanging air intake.

Air-conditioning is not really a plus. In addition to the initial cost of the unit, air-conditioning consumes large amounts of energy. Older houses were designed to be kept cool before air-conditioning was around. Deciduous trees on the south side, awnings, and window shades help prevent the sun from heating up the house. Operating windows with screens allow the breeze to help keep things cool.

Be especially aware before looking over a house of any special equipment you might need to meet health codes or licensure standards. If you are going to be required to install an interconnected fire alarm or a institutional range hood, you ought to know the costs.

Moving In

Once you have selected a house and have money committed, more than likely, the biggest problem will then be the time it takes to decide on renovations required by codes and standards, to make those changes, arrange for inspections and approvals, and do the paperwork.

3-25. Heating fuel is costly, so that checking energy efficiency is important.

3-26. Gaps between the window casing and the wall are prime culprits in infiltration.

3-27. New wiring is usually installed in a metal conduit for safety.

3-28. Water heater capacity is an important consideration, as is the energy efficiency of the unit.

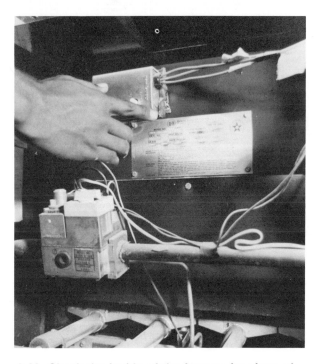

3-29. Check the inside of the furnace for signs of rust and other neglect.

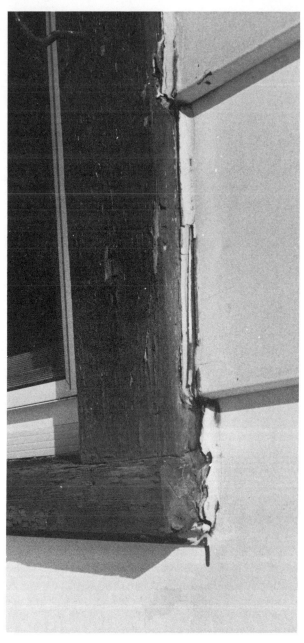

3-30. Caulking would solve the infiltration problems that these gaps cause.

Try to involve the people who will be living in the home in this process. As much as you possibly can, defer nonessential decisions for group discussion while concentrating on those things that are necessary to get the house opened. Such things as interior painting and color selection; furniture selection, purchase, and arrangement; all decorating; assignment of room uses; some drywalling (but not taping); floor refinishing; and weather stripping can all be purposefully deferred. These are all things that can be done by amateurs, depending on their skill.

Almost everyone remembers being part of starting something new. It was a time of bootstrapping, of relying on unknown, untrained abilities. It was hard but exciting work. This spirit is to be encouraged in opening the house. One home recently sponsored a "better bathroom" contest, allowing residents to propose decorating schemes for the bathroom, with judging by a local architect and a cash prize for the winner. The result was both a better bathroom and more responsibility by residents in caring for the bathroom.

There are two basic approaches to moving into a new house: by doing so quietly or by first laying the groundwork in the rest of the community. Some agencies feel that if they move the group in with "trumpets blaring" they will meet a lot of resistance. And although such an approach may be motivated by real desire to educate the public and dispel misconception about group homes, it is sometimes ". . . at the expense of the client," according to one operator. "If you're going to get people out of the hospital, do it as quick and as clean as you can. You don't have to lie to anybody or fool anybody, but at the same time you needn't volunteer information. I worked in an institution long enough to know that if you start to ask for permission to do something, you're in trouble. The best thing you can do is just do it and see if anyone gets angry

about it. And the longer you can do it and the better you make it work, the less likely they are to get angry. If they never find out about it, that's all right. When we moved in, we did a lot to let neighbors know that we weren't going to be a burden on their already limited resources. We made it clear that we had a business and everyone worked. But beyond that, we just moved in. It's not normal for anyone else to go through the neighborhood and ask permission to come in. We didn't either."

On the other hand, working openly with the community can sometimes assure later cooperation. On group began by ". . . gradually building community support by involving community leaders at every level. We formed an ad hoc group that includes people from city council, clergy, leaders, and so on; the sorts of people whose influence can make a difference. And we've been very careful to make it clear that this was not just a mental health thing, but the concern of the whole community. So, for instance, we don't hold our meetings here in the mental health board's offices. Now we're beginning to get a lot of cooperation from the community. I no longer shudder when I know something concerning the home is going to come before the zoning board. They know more about us now—what we are trying to do."

Another operator felt that "talking with the neighbors ahead of time helped. They had a lot of misconceptions about our clients, assuming they were all a threat. We were able to assure people that their property values wouldn't go down and that the neighborhood would probably be safer with our group in residence."

There is no clear answer to this question of moving in quietly or not. It can depend on your own temperament, that of the local zoning board and public officials, the past experience of the community and neighbors or the phase of the moon. Both points of view have their advocates.

HOUSTORY ONE:

THE NEAR WEST LODGE

Bill Whan spent many years working in state mental hospitals in a variety of capacities, from ward staff to unit director. He finally moved into the community after first forming—from within the hospital—a group of twelve long-term chronic patients. Together they started Near West Lodge, named for the Cleveland neighborhood in which it is located. With a state grant, they purchased a small one-story building four blocks from the lodge as a site for the furniture refinishing and manufacturing business now operated by lodge members.

Our most valuable thing was a realtor's listing book for the side of town we were interested in. These books have hundreds of photographs of houses and coded descriptions for each. It was something you could pick up and hold in your hands and talk about. Kind of a wish-book catalog. We kept it on the table in the big office room at the hospital, which was an open place where people hung out and talked about the future. Once or twice a week we'd get a carload of folks together and drive out to look at places. We cruised the neighborhoods and looked at everything—we saw some real losers, and some beauties. Later we'd get appointments to visit the likely ones. A lot of places we could eliminate out of hand like this, and we had lots of little short affairs with several dozen others.

With the place we finally found there were a lot of lucky factors—that's not to say that sheer persistence didn't count in the end, though. We looked for nearly a year. As it finally happened, we had seen a street, a neighborhood that looked just right after a while, and we had said, "I like this street. Wouldn't it be great if one of those big houses went on the market?," and a couple of weeks later we saw the house in the listings.

We went for a visit and it was occupied. In fact, it was a rooming house. The owners, a nice old couple, were living there. We liked them from the beginning and were upfront with our plans. We were overjoyed to hear them say, "That's wonderful, just what I was hoping this building could be used for." We walked into a tradition. The owners told us the house had been taking in roomers since 1905. It was a lucky find. Any find is a lucky find, but I think the luck is always there.

We had to find out what needed to be done to make it habitable and what needed to be done to make it legal; those are two different things of course. And we started to think about the neighbors. The owners had lived there for thirty years and were very good about helping us to learn about the neighborhood and who lived there.

The neighborhood itself was kind of a hodgepodge. There were some gracious old Victorian houses that had been restored, some that had been broken into apartments and some that had fallen into disrepair. Absentee landlords. Not too far from some welfare hotels; not too far from some expensive and trendy restoration. Downtown was quite close.

After the owners moved out, I moved in myself for a month and learned a lot. Sometimes in this business you have to stop and take a breath and ask yourself what's normal. It's all part of the difficult transition from the institution. It's sad that you have to remind yourself to do this, but you have to.

For example, when I was living in the house, I went next door to borrow a rake, and chatted, but it would have seemed phony to talk about mental health. We decided it's not normal to go up and down the street, knocking on doors, saying "Is it all right if I move in?" So, after we had closed the deal on the house and signed all the papers, we did go around and talk to people we had learned were community leaders. We were convinced also, that if you're going to have a group home in a community, you should be doing something for that community. You start in a sort of one-down situation, and you have to persuade folks that you're not just taking all the time, that you're going to be contributing too. You shouldn't be neutral; you should be active.

We made it very clear we'd be doing a lot of things: running a business, creating jobs, taking care of the neighborhood and not just the front lawn, buying things locally, residents joining the block club, and so on. But, as we expected, the block club finally gave us a call. "We hear you bought the so-and-so house and you're gonna start one of them group homes." So we came to their next meeting; it was their best-attended meeting ever, and even though we knew full well why so many people were there, we were careful to praise the club for its high spirits and sense of involvement.

Any time you meet like this with folks in the community, particularly with people who are potentially hostile, it's invaluable to have at the meeting the people who are going to be living in the house. The neighbors are going to want to talk about "those people," and there's no substitute for having "those people" sitting right there taking part in the meeting. There's a big difference between *those* people and *you* people. One of our guys talked at length about his troubles from having been in Viet Nam, and everyone quickly said, "Oh, we didn't mean you, we meant those people." Another guy fell asleep halfway through, and I think they started to realize we might be kind of boring but we weren't very dangerous. People are afraid of people they don't know. It was fun to watch our guys asking the neighbors in a worried way about crime on the street.

After we got that fearful kind of ignorance out of the way, people wanted to know about the program and the kind of work people were going to do. I don't want to give the impression that the meeting ended in sweetness and light. A lot of folks grumbled and left early, particularly the ringleaders of the antilodge faction, but everyone else could see they were a sour-grapes sort of minor faction after all. This minority wanted especially to know if we had closed the deal on the house and gotten our occupancy permit from the city. That was where they wanted to intercede, in zoning and official kinds of ways, but we made it clear that all that was taken care of.

We didn't expect any trouble along these lines because the house was already being used in exactly the way we were planning, although we were going to have to develop more common areas and a communal kitchen, features the house didn't have. We knew we'd have to do a number of things to meet various codes and requirements. There were so many authorities to satisfy after we met our own criteria about what a good house was; the agency that accredited us, the state Department of Mental Health, the hospital, and several city departments.

The tactic we found most helpful with all the inspectors who actually visited the house was to treat them all like helpful experts. After all, they are experts, and treating them that way makes it easier for them to be helpful. We asked all kinds of technical advice about how to meet various codes, and the changes we made became a kind of joint project between us and the inspectors, so they had a kind of personal pride in the results.

On the other hand, we spent a lot of time down at city hall with some real deadheads who were unwilling to listen and never even visited the house. We'd have to start from square one at every encounter with these bureaucrats. They'd say, "These people are mental aren't they? Then it has to be an institution." It took a long time to resolve the problems, but we didn't solve them through rational discourse or through playing the system as it exists on paper; we solved them by political maneuvering. We wanted to avoid a hearing before the zoning board because the opposition would have shown up in force. The zoning board, we were sure, would have said, "This looks like trouble," and the impulse is always to not take risks in a situation like that. So we went to our board of trustees and our friends in the neighborhood and started calling in favors. We called on politicians. I remember crashing a birthday party for a city official, given by his office staff, and cornering him in the back room with a cup of brandy in his hand to enlist his support. We slid through. We were never sure just which maneuver won for us, there were so many of them going on.

A lot of the evils we ran into were the evils of abstraction. The generalizations people made, the flow charts and tables and graphs. Stereotypes and statistics. Those were the kinds of poisonous things that got in everybody's way. They're self-serving and reproduce themselves endlessly. They give lots of people something to do, but these people never get to meet the people their restrictions influence. The goals people have for their own lives get lost because of this, their chances to meet their goals, the kinds of human relationships between people, the chance to take chances.

Too often the people whose job it is to make sure we get quality in services are only interested in quantity, in numbers and properly filled-out forms. When you stand so far away from where things are going on, sometimes the only things that filter through are the numbers and the things you can write down on paper and feed to computers, and things go wrong.

If you're in the group-home business or the mental health business, it's your business to take risks. You have to. If you don't, you might as well sterilize everybody. There's nothing more routine than the hospital; there's nothing more regulated, nothing more safe and secure, so why shouldn't people just stay there if absolute lack of risk is what you're looking for. If you don't know that life means change, and change means risk, you're in the wrong line of work. If things are going to become more normal and more human, they've got to get looser. They've got to get less prescribed. A prescription is what the doctor gives you. If everything in your life becomes prescribed—the shape of your room, how many stairs are outside—your whole life might as well be an Rx. You're not going to grow any.

Rosemary Kulow and Jim Buskirk are residents of Near West Lodge. After hearing Bill Whan's houstory, we turned to them to get another view and began by asking them what they would tell someone who was just starting a group home.

Rosemary: Make sure it's convenient to shop.

Jim: Don't be across the street from a bar. I have the third floor bedroom with the window facing right there. You get a lot of noise at night with people coming out of the bar. That's the first thing they tell you when you move in, "If you like to drink, fine, but don't go to the Clinton bar cause it's bad news." They don't want any bad feelings between that bar and this place. If someone goes over there and gets in a fight or somebody gets mad at them, you don't know what those people might come back and try to do. Set fire to the house, you never know.

What about other neighbors?

Jim: They keep pretty much to themselves. I think they don't realize that this is a group home, they never mention anything to that effect. I run into them every once in a while when I'm walking around, coming home from school. They're friendly. They say "Hello" and'll be real nice and everything.

This neighborhood is good because things are improving, being rebuilt, redone. But for some reason you still have a lot of, they're not low-lifes, but destitute people living here and they can make your life miserable. They hang around the street, hitting you up for money, cigarettes, whatever. I'm trying to make my own life go, I don't have time to be charitable to everyone else who walks up and down the street all the time.

It's not dangerous it's just that there's people always bugging you for money or this or that. There were people, I guess, who lived here who had problems over there. Everybody I've talked to said, "Don't get off on 25th street if you're coming home late at night. Get off downtown and catch a bus back up Detroit Avenue. That way you avoid the bad section."

But you work everyday on 25th?

Jim: During the day it's fine, but at night forget it. You don't want to be caught . . .

Rosemary: But the hospital is right there on the corner of Franklin and 25th, and the emergency room is right there. I'm not afraid, if there's something wrong and I have to go to the hospital. And the Y, in fact, is right up the street. And a block north is Detroit Avenue to catch the 26 bus line to downtown. It's convenient.

How about banks, post office, that sort of thing?

Jim: There's a bank right on 25th [a 10-minute walk] for cashing checks from work.

Rosemary: There's a money exchange right on 25th too. The closest post office is downtown, though [20-minute walk]. There's churches right around, church services and masses. There's St. Malachi's, and an Episcopalian church and a Unity church or whatever right on Vestry.

Are they walking distance?

Jim: Yeah, I'd say at most it's a 10-minute walk.

What about groceries and shopping?

Rosemary: Most of the house shopping is done at the supermarket up on Lorain. Some of the bulk shopping is done at the Food Co-op and food bank.

What if you need a pack of cigarettes or bottle of pop?

Rosemary: Sam's is right down there on 29th and there are lots of little stores around.

Jim: I don't like the local stores 'cause they can charge whatever they want. It's not like a Convenient or a Lawsons that are open twenty-four hours. There's nothing like that close by. It's outrageous what they charge at these mom-and-pop places. And their hours are strange. If I run out of cigarettes I have to wait till morning.

What about fast-food places or movie theatres?

Rosemary: There's a pizza place around here somewhere.

Jim: When I'm on lunch over at work I go to a little place called the Liberty Restaurant. It's right across the street. It's a little dinky place but for what you pay, it's worth it. The food's good, which is surprising.

Rosemary: We used to go to the movies every week. But then nobody wanted to go, so George, our activities director, said to let him know whenever we want to go. He takes us out to the malls. We use the workshop van.

Jim: The malls have four or five movies showing, and you have a choice if you don't want to see what someone else wants to see.

What about other transportation?

Jim: Easy access is important. You are on your own here; what you do with your free time is up to you. Of course if you have to walk two miles to catch a bus, you're not gonna go anywhere. Public transportation is very easy here 'cause you're so close to downtown. The unfortunate thing is you have to stay off certain streets around here at night. They're too dangerous after 11:00.

I go downtown a lot and use the library, and I go on weekends and visit friends that work at a bookstore. It takes me five minutes, I can walk in fifteen minutes if I want to.

Rosemary: I visit relatives mostly. It seems like everyone here just stays tucked away in their rooms most of the time

How about work?

Rosemary: I enjoy it. I'm secretary over at the office. I like the responsibility, something to do.

Jim: I like my job [in the furniture refinishing shop], but right now I'm trying to do too many things at once. I just started at community college, and I haven't gone to school for six years and wasn't prepared for all the studying. Then my job at work and jobs around the house here.

Rosemary: The home should be accessible to schools, libraries, places to study at night. There are people who want to go back to school.

Jim: What's nice here is you're living on your own. You don't have family supporting you, but there are people around of your own age you can sit and talk to. You don't have to sit in your room being bored. There's always somebody here, even when I come home late from school at night, there's usually at least one person in the kitchen.

There are people here if you want but you can get away to your own room. You're not all crammed together all the time. That's what I like about this place, individual privacy. There's no one standing over you saying do this, do that. They expect you to take care of yourself. You have to do that when you get out on your own.

HOUSTORY TWO:

THE WELCH HOUSE

*C*harles and Suzanne Welch are professional group-home operators —professional in the sense that their home is both their source of income and also the center of their lives. They raised their family and made a way of life amidst the difficulties and freedoms of running a group home. We chose to focus the interview on the problems that arise when dealing with the many regulations and inspectors encountered in the operation of group homes.

Their pre–Civil War mansion in Portsmouth, Ohio, is the home they always wanted. Owning it was made possible by operating a group home and having an income from fourteen residents. As they learned, the house had been through many incarnations before they owned it.

Suzanne: I've gone to some of the oldest people in town, and they can remember way back when Mr. Lawson owned it. He used to ride his big white stallion around the yard. But no one's left who can remember way back to the 1850s.

We got involved through the VA [Veterans Administration], oh, fifteen years ago. They called them family-care homes. I went to the state in 1978—I heard they were interested—and we got to be a pilot. We were the first large extramural care home in this area. We have both VA and state psychiatric clients living here now.

Charles: The codes are what tore us up here. See, this operated as a nursing home for years. It had been a nursing home under two owners, and what we were looking for was a big house, so it would meet the codes. There'd been some hanky-panky somewhere along the line, and it didn't meet codes. The wiring, the plumbing didn't meet codes. The stairwell and this and that.

They passed an ordinance that whenever a piece of property changed owners, or usage, all the inspectors would automatically come in. And it just so happened that we were lucky enough to be one of the first ones. And boy did they hit us hard. All that's been dropped now. We have an $1100 range hood in there that no one's ever checked on.

Suzanne: They insisted we had to have one and wouldn't let us open until we did. But no one's ever come back to see that it's installed.

What code classification did they put you under here? It's obviously not a nursing home any more.

Charles: Well, that's another touchy one. They didn't quite know how to classify us. We're classified as a rooming house in Portsmouth. But the rest of it was just based on what we intended to use it for. They made up the rules as they went along.

Suzanne: And you mentioned mental illness or something like that, and right away the fire inspector went right off the ground. He said, "Oh! retarded people!" And we said, "No, they are not retarded, they're psychiatric clients. Mental illness, not retardation." He said, "Same thing!" My husband said, "No, you could be one of our clients one day. That's the difference. You're not retarded, but you could be one of our clients." "Oh!" He went back and wrote a recommendation that my husband not be allowed to operate anything!

Charles: I was considered a fire hazard because I questioned their methods for calculating the sprinkler system. I asked about the calculation for the number of heads and the supply pipe and this and that. It comes out a hundred percent—every head activated. And we'd have to have a supply pipe to carry the volume and the pressure. I said that's ridiculous. If there's a fire in this house that will activate every head, the house is gone. Everyone in it is either already dead or outside or whatever. Well, he didn't like me questioning. He sent a letter more or less calling me a fire hazard. After all this, they didn't even bother to check the fire alarm we do have.

What type is it?

Charles: It's a Notifier. It has four smoke detectors and four bells, with four pull boxes. The main unit is in the basement. But they didn't even check that. Then they laid this stuff on us about an $1100 range hood. There are restaurants in town that operate without that. He said within a year everyone would have it.

How long ago was that?

Charles: Four years ago.

Does every restaurant now have it?

Charles: No.

Did you ever have the feeling that they were purposely trying to make it difficult for you because of the people you were working with?

Charles: Yeah. Sure do. Oh, at one time, bein' friendly with the building inspector helped. The electrical inspector was the one we had the hassle with. See . . . he's been an inspector close to 15 years now. He came in and inspected and saw there was no conduit. Now it was kind of a cover-up job down[stairs] here—as far as conduit, they had some of it exposed and whatever, just to look good. But the second and third floor didn't have it. I said, "Bill, haven't you ever been in this house?" He said, "Yeah . . . I inspected the fire alarm system." I said, "You didn't . . . check out the wiring at the time?" "Well . . . there were a few gray areas there . . . with the state . . ."

During that time were there people living here?

Suzanne: No . . . they held us up. We were seven months when we anticipated two months.

Did you have any vibrations from the neighbors?

Charles: The feedback we got was a lot of people were glad someone moved in here . . . this big old house vacant. A lot of 'em are older here. They were happy to see us come in and see lights on in the place. We do the neighbors back here a favor by watching the buildings. There's someone here twenty-four hours a day.

It's the local inspectors that are the ones that give you the hassle, and depending on the neighborhood. Now . . . there are places in this city where you wouldn't want to try to start a group home like this. If you do it in this end of town, you're all right. But you move above Offner Street or out beyond 17th . . . 16th or 17th . . . forget it.

Suzanne: By the graveyard . . .

Charles: To do something like this you have to be real choosy where you go.

Suzanne: Up on the hilltop, there, they sneaked one in. Young kids . . .

Charles: If you could just . . . go ahead and do it, seems like you'd be okay. Don't advertise, don't go to any of the inspectors or anything, just do it. You'd seem to be all right. But we tried to do it the right way. Got approval from everyone.

The main thing goin' into this I can see is to have some money . . . 'cause in four years with the income, we've barely been able to keep this place up, let alone make the improvement. You can see it's a gradual process.

Suzanne: We go slow.

Charles: This house has twelve-foot ceilings. Every time you do anything, it seems to cost about double a normal house, and about three times the work.

Suzanne: Last summer was when we were supposed to start the roof. We had a hailstorm. So we just went on vacation. It was too much. But my sister-in-law, bless her heart, came in to cover, so it worked out. It's the only way you can do it.

Charles: The gas budget here you wouldn't believe. The gas company likes the place so much they're soon gonna own it.

Suzanne: They just increased the amount on the budget by $85. It's now $375 a month.

Charles: The increase is more than most people's whole budget!

Suzanne: Sometimes I'll get frustrated 'cause there's things that need to get done right now, or you plan to fix somethin', and somethin' else comes up and you have to do it, and you can't do what you were thinking.

Charles: There's a problem here with the health department. They drop in here whenever they feel like it. Now the way it is is the way it has to be. We can't change anything. We can't even fix that counter without presenting a plan so it's approved. Otherwise we'd be remodeling. It's under a grandfather clause kind of deal where it's been this way so long, it has to stay. We have a foodservice license which costs $37 a year. I forget what the rooming house permit is—oh . . . that's just $10. That range hood sprinkling system would cost us $65 twice a year to have it inspected. The guy more or less walks in, looks at it, and walks out.

When we started here four years ago, this fire alarm system wasn't required. Well, the state came along—was it last year?—they contacted us and said, "Bad news. You've got to have an interconnected fire alarm system." I said, "We have one." "Well, we didn't know that." They'd been here and inspected every year. So they wanted to come down and see it. Well, they inspected it. Would you believe this? At the time they came it was inoperable. It had burned a transformer out. So I called—oh my God—all over the place . . . Lincoln, Nebraska . . . trying to get a transformer for the thing. I finally found one. Puts out eighteen volts, costs $750. I think it's a 3-by-3 transformer. It took maybe three weeks to find it and another six weeks to get it. In the meantime, we got to have the thing or go out of business. So I started checking around with the local installers on what it would cost to have a new one put in here. It was $5000 for an equivalent system. I said, "How much for just the system. I'll install it myself." He said, "You're not allowed to; you have to be certified." So that's why I say to get into this business, you better have some money.

Suzanne: Down the road we'll be nonexistent. We operate on a per diem basis . . . per head. If we have a vacancy, that's our problem.

Charles: Here's something we have found out. You have got to get away from it. It would be nice if we could take a weekend . . . at least a weekend a month and just go . . . somewhere. It can literally drive you nuts. We get to arguing over things that don't mean a thing. All it is is pressure. There's fourteen people with fourteen problems. And they compound these problems and bring them to us. If anyone would ever ask, that to me is the most important thing . . .

Suzanne: Your time . . .

Charles: You have to have time to yourself.

Suzanne: . . . and privacy.

Despite their realistic view of the difficulties in running a group home, they also love what they do.

Charles: It makes you feel like you're doing something right.

Suzanne: You're never lonely here. You can live in your home and not have to go out and work. You don't have to answer . . . well, you always have to answer to people. But not in the sense that you've got to punch a time clock. If I feel like I want to go downtown in the middle of the afternoon, I may have to double time when I get back, but it's my choice.

Charles: We've learned to talk straight. There's no doubt about what was said when you talk straight. But they [the residents] think they're supposed to hold it all in. After awhile, you can see something's going wrong, and you have to talk. Well. It kind of makes you feel good, walk away, look at that guy, and he's got a smile on his face . . .

Suzanne: They don't trust people. I can see why. I've had 'em say to me, "You gotta watch everything you say." You'll talk to one doctor, and you'll hear the doctor . . . give the same questions. The guy gives the same answers. They know the routine. They don't want the doctor to know there's anything wrong so they might get stuck in the hospital. Here they listen up, and they talk . . .

Charles: We have found out things about these guys that, twenty years in the hospital, they have never found out. We live with 'em twenty-four hours a day. After awhile they get to trust you. They'll tell you something and say don't tell the doctor. You better hadn't. They don't talk to you again.

Suzanne: You just try to talk them into it . . . it's something they need to work out . . . then do discuss it with this person.

Charles: If you think it's serious enough, you try to talk them into tellin' the doctor.

Suzanne: Here they see us relate. Now, that's where having kids helps, because they see us relating and going through the same things that they went through as a kid—our kids go through it. And I'll tell 'em why I do it. I love my children, but I can't tolerate some kinds of things. They just learn through watchin' us as a family, go back and forth . . . and they can learn to overlook some of their family.

Charles: We've never had any formal training. We're just more or less goin' along and feeling it out. Thought about goin' to school and whatever, but just might get whatever it is we have schooled out of us. It's been known to happen. I look around at these psychiatrists, and something went wrong somewhere.

Suzanne: We talk blunt. Straight. They understand that. We all understand that. Around here we're not all connin' each other all the time.

Charles: It's hard to imagine getting up in the morning and going to a 9-to-5 job. We've been together ever since we've been married. All those years we've been together doing it ourselves. I'd feel lost without having her around seeing to it I'm doing things right.

One fella said, "This is the craziest damn place I've ever lived." "You like it?" "Yeah, I like it!"

HOUSTORY THREE:
181 HOUSE

The 181 House was especially interesting to us because of its community: Tiffin, Ohio, a small industry and college town that has recently experienced the closing of an old established state mental hospital. We acknowledge that we are giving 181 House's program and staff short shrift in order to emphasize the importance of the community support it has received.

Our conversation was with Roger Murray, coordinator of the Ohio Department of Mental Health's District 5; with Charles Curie, the executive director of the agency that operates 181 House; and with Tina Pine, the director of residential services. We started by talking about community resistance.

Charles: The basic underlying principle is having contact with the neighbors and the community on a regular basis. That's kind of risky—if you know people are hostile, you tend not to talk with them. We went about it informally and did a lot of talking over the hedge. Our neighbor had no problem with our playing volleyball—it was just that the ball went into her garden. So one of the staff people got involved, and we moved the volleyball area a little bit over, and she saw we were willing to work with her. Now she's gotten interested in us and brings vegetables over from her garden.

Tina: Sometimes there were bitter feelings from the neighbors in regard to mental health clients living in such a nice facility. They look out their back windows and see clients sitting in a very nice living room with color TV and air conditioning. And they just couldn't understand that. But I think it was because they thought these people shouldn't occupy this nice housing.

Charles: It's helped, too, our effort to support and cooperate with the police department—law enforcement—these types of agencies in town. We don't run to them all the time when we can handle the problem ourselves. And when they have a psychiatric emergency, we help them handle it. So we've found that people in those agencies defend us to the community and support us, and that can be very critical.

Roger: Some of the resistance in this community originated because we were tearing down buildings on the grounds of a state hospital and at the same time building new facilities in the community: "Why do we have to build new when we could convert a farmhouse, or why do we have to be so close to town?"

Charles: The location has worked very well for residents because they have access to services in town.

Roger: The neighbors wanted assurances that this was not damaging their property values or physically intimidating their children or their families.
  In Lucas County, where the state Department of Mental Health helped purchase houses, there was tremendous resistance, and they went the route of getting advisory groups made up of neighbors that the managers of the group homes met with. They dealt with issues like: "How do you know that their kids won't sell marijuana to our kids?" Of course they responded, "How do you know *your* kids won't sell marijuana to our kids?" But once they felt their security needs were met, these people were very supportive. Once they felt there was someone who could be held responsible.

Charles: I don't think the neighbors or the community have ever indicated that they wanted more input than they've already had. You could end up getting counseled and advisoried to death after a while, too . . .

Roger: Tiffin is a very caring community. It experienced the closing of a major state hospital. That was like a shock. Most everyone knows someone who worked in the state hospital—every neighborhood in Tiffin. And these former state hospital employees are familiar with clients and

Charles: concerned about the adequacy of programs. We're especially aware of this when it comes time to pass another mental health levy!

Charles: We have an affiliation agreement with the Toledo Mental Health Center. A fairly in-depth contract delineating the responsibilities and roles of agency and hospital. It clarifies things. The staff at the center knows what's expected. The staff here knows what's expected. Having that spelled out, I think, averts a lot of things.

Roger: Also, a hospital–community service contract was developed from the Huron County contract and then somewhat modified for the Sandusky Valley area. It's probably the most extensive contract between a mental health board and the state hospital. It delineates mutual expectations and a process of information-sharing which is crucial to planning and service delivery.

Charles: I think the critical thing is the liaison workers. They've done a lot. Their responsibility is to check, on paper, on everybody from the three-county area who's in the state hospital. If we had no liaison workers, I'm sure we'd be getting phone calls: "Hey, we're ready to release this guy, do you have a bed?" And we'd say "... yeah." And then they'd send him down on a bus ... (laughs) ... and then on our doorstep we have a violent, acting-out client that we're not prepared to meet.

Roger: In some programs you may have an avant-garde aftercare worker who may have a lot of compassion in a situation where a person is very disturbed but has very little understanding of what dynamics might occur. Bringing this person into a situation ends up upsetting an entire residential facility. And in turn, that can spill over into the community, and it can take years to repair that.

In addition to the hospital–community service contract and the agency-hospital affiliation agreement, it's the board-provider contract that ensures that the agencies develop programs that meet the needs (prioritized by the county board), and the board takes a hands-off approach to the agencies. The board will tell the agencies who to serve and what kinds of results they want and will go to an independent monitoring to see that it was in fact done. The agency people run their own ship and do the best job they can do and work together with the board from there. But the principal concern, and it has been, is getting service to the clients. That is the bottom line.

HOUSTORY FOUR:

THE CINCINNATI LODGE

We were first interested in the Cincinnati, Ohio, "Lodge" because it successfully involved residents before the house was even officially opened. When we visited, we were struck by the difficulty we had distinguishing staff members from lodge members. Everybody seemed to have a vital interest in things. Lodge members operate their own landscaping business in addition to operating the lodge.

Sue Powell, a member, and David Whetsel, program coordinator, began by talking with us about their decision-making style.

David: If it's a group decision—even a wrong group decision—it's still better than a right decision made by me. That's the main thing we do around here—group decisions. That's vital in any group—understanding that it is a home. In a home not just one person makes decisions. Everybody does.

It would be a lot easier to assume all the control and make all the decisions. It's a lot harder to have everyone making decisions, because you leave yourself open for discussion, arguments, people being angry—you leave yourself open to being wrong. Self-confidence comes from making decisions—right, wrong, or indifferent. And if the decision is wrong, then you deal with the fact that it's the wrong decision, and you change it like anybody does.

Sue: Personally, I feel a person should make their own decisions, and if they make the wrong decision, then it's how you deal with this.

So it takes more time?

David: Oh sure. Every decision being made is up for discussion, arguments, criticism, whatever. House decisions, business decisions, decisions that affect everybody have to be made as a group. If you're a control person, basically you're in control. If what you say is wrong, then you deal with that . . . you change it. It's quick and it's easy and it's done. Where, if you have fifteen people, one decision may take an hour; where you make it on your own, it takes three minutes.

Sue: In the hospital you have everybody telling you do this, do that. They're basically making the decision for you—you aren't making it. You've got to be in control of your own life. There's not always going to be someone there to make decisions for you. You feel much better about yourself when you're doing things on your own. I do much better living with people than on my own. We sit around once a week making out a menu. Everybody has an input into it. And then we decide, you know, who's going to cook those days.

What if nobody wants to cook?

David: Nobody eats.

Sue: Well . . . somebody fills in for the other person.

David: We had one person who was put on probation by everybody. And then taken off probation by everybody.

Sue: He had trouble getting up in the morning. He wasn't around when he was needed.

How about your own room?

Sue: My roommate and I, we keep our room together. She has one part of the room, and I have the other part. She's doing a lot of moving [furniture] around, more than I am. We painted our room, blue, it's my favorite color.

You picked the color?

Sue: Yes, we did. There was leftover paint here.

David: It went from beige to blue.

Sue: We made one area like a sitting area. It has a couch and a stereo, television. I used my own dresser.

Saturday, we ran a yard sale here. We sold odds and ends of stuff . . . and did pretty good. We could've done better, but I feel it was pretty good for a first time.

David: The neat thing is that all the stuff that was sold was theirs—it was all individual, personal stuff. And the staff wasn't involved. The only thing the staff did was to occasionally help price some stuff. And the money they made went to buy a couple of tents in order to go camping next week. No lodge budget was spent, not a dime . . . except for the ad in the paper, and they paid that back.

How do you handle housework?

Sue: We do our own wash and keep our rooms clean.

Do you have room inspections?

Sue: Yes.

David: We are like landlords. Every so often we have an apartment check.

When we moved in to the house, it was vacant. There was no paint on the walls . . . half of it was drywalled and half of it wasn't. We tore down walls. Pulled out partitions. Put in a couple of windows that weren't there where air conditioners were. Basically we took it and did exactly what we wanted it to look like. The members here are the ones that did 80 percent of the work.

How did you decide what you wanted it to look like?

David: Basically it was pretty good. Each room had its own bath, and there were some open spaces. We're running a business here so when people walk in the front door we want them to see the office. Basically we walked into the front door and said, what do we want people to see if they were coming in to hire us?

Who was involved in that decision?

David: Myself and another work supervisor and seven members who came from Longview State Hospital for a day program for four weeks.

Was that by accident?

David: That was planned. They'd been working out at Longview, I think, for almost a year. Part of the plan for moving into the lodge. They started out with, I think, fifteen people. When it was time to move in, they were down to like seven.

You had to make deals with the building inspectors to occupy the house before it was actually up to code, right?

David: Oh yeah. There were several things. There was something about the kitchen where you could only serve up to six people or something in order for us to pass regulations while we were still working on it. So staff couldn't eat. Made it kind of rough (laughs). We couldn't move on to the second floor [either].

Why?

David: There was a window that didn't meet regulations. It opened out and had a bar down the middle so you couldn't climb . . . jump out.

As a fire exit?

David: Right.

Did you talk with the inspectors? How was that?

David: I tell you, they're very lenient . . . if you work with them. I mean, if you approach them in an abrupt manner, you get absolutely nowhere. Basically, the best thing I found was do things when they tell you to do them. Don't put it off. And if they come back the next day and it's done, they're impressed and they're more than willing to go the extra mile with you. But if you put it off, then they get upset. They're on a time schedule, that's when they start to get defensive . . . or they don't appreciate it. They give you a list. It's a real concrete list. They say this, this, this. You either do it and you're covered, or you don't and you take all the backlash.

The things they ask . . . the majority of the things they say to change are not major. Most of them are things like get your refrigerator on wheels so it's sitting off the ground. You know. Make sure this piece along the door is nailed down better. And when they come back through, all they do is go through the list.

Sue: I like the way the house is made. We've got one big break area downstairs, and it's just a homey place—it's just a nice place to live. It's a clean, neat place to live.

David: I think the main thing is the personal space. You can go to your room and have a key and the door locks . . . they have a key to their room and to the front door . . . and nobody can get in if they don't want them to get in.

Sue: It's a safe place.

David: I think it's the work that's been put into it. The willingness to put in that work . . . makes it a home.

You put a lot into it, Sue?

Sue: Right . . . I feel we all put a lot into it.

Why are you willing to put so much into it? Is David such an ogre?

Sue: No! (laughs) It's like maybe they haven't ever lived in a kind of place like this.

David: This is their place. That's the feeling they have.

Sue: Right. It's like one big family.

David: And I think we have to come back to the basics again. They make the decisions. When you make the decisions about everything that's going on around you, then you're part of that. The building's part of that. You feel like it's your place. You take care of it. When they make decisions for you, then it's their place, not yours. Does that make sense?

Sue: Right, it does.

Going back to earlier questions, what makes it more difficult?

David: I explain everything. With rent, I don't just go and say, this is what you owe, and that's it. If I can give some specifics, I think it will help a little bit. (This month we started a new system so that rent depends on what you earn.) By explaining the whole system, they then were able to argue and give me their gripes and complaints . . . they were actually real mad because they understood the system and because they saw flaws in it. If I would have come in and said, this is how much you owe, then how can they argue . . . how can you have any problems with that when you don't know anything about it? Basically, I showed on paper the whole system. How the rent was figured, how much they made on the job, and how many hours they worked . . . the whole thing. I guess explaining enough detail to people gives them enough information that they can fight back . . . and say, well that's too much or how come I'm not getting this.

If a decision has been made simply because the executive director says it has to happen, then they deserve to know that [too]. If they do know that, they can say well that's okay or can we fight it? Our goal in the next few years is to have less staff.

Sue   I'm excited about it. I feel that we will get someplace. It's gonna take time and a lot of hard work, but I really feel we'll get there.

David: I don't think anybody questions that. There's so much energy and everyone feels so positive . . . I don't think anybody really questions the fact that we're going to get what we're going towards.

The whole philosophy is give them more responsibility and they will accept more. They are.

Also, it might help you to know that we just got fired from our first job. The lodge got canned . . . which was a real shame. It was a mowing job and we couldn't get it done in two days like they wanted. We had regular mowers and it was five acres of land. It took us about three and one-half days, and they wanted it done in two. The quality of work was fine . . . they had no problem with that.

Sue: I really don't think it was our fault. I really don't. How did we deal with it? We talked about it, and a lot of people were really disgusted, and we had some really bad feelings about the place . . . 'cause I really feel that they done us wrong.

David: Normal feelings.

Sue: They handled it the wrong way. The way I felt about it is if they had wanted it done in a certain amount of time, they should have written it up in the contract.

David: We're learning about bidding and contracts . . . I would like the community to look at us as, instead of their taxes paying us to keep this place going, that someday the members' taxes here will be paying to keep them going.

HOUSTORY FIVE:
PANTA RHEI

*A*t the time of this interview, Claudia Reiter was the residential program coordinator at Cleveland's Panta Rhei. (Panta rhei, a Greek phrase, means "ever flowing.") The lodge operated by Panta Rhei was one of the program sites visited in our research study. Originally these were two houses side-by-side; a major remodeling that physically joined the two houses with a new addition had just been completed. The newly remodeled facility reopened with a new group of residents who, typically, had experienced long-term or repeated hospitalizations for mental disabilities.

You have to be willing to let things look like they are going to fail a lot. The clock's ticking, and no one is making dinner. People are there, waiting for you to take over. I keep saying, "I'm not making the meal, you are." It's hard not to say, "Okay, let's get going, everyone is going to be hungry." You have to stop yourself from doing that. You can't run things. You have to live with the fear that it will all get out of hand. It really requires a different style of doing things. And Judi [a coworker] and I are still involved in the learning process of how to do that.

The amazing thing is that one of three things happens. Whoever is supposed to cook shows up and throws it together in a big hurry or makes something else that's quicker, or someone else who's not supposed to cook says, "I saw them downtown and know they won't make it so I will." Or nobody cooks, and everyone wanders in and makes a sandwich, sometimes grumbling, sometimes not. It's interesting that people don't give each other more of a hard time about it. Maybe later [after being together longer as a group], they'll feel they have more of a right.

But to me it's really exciting. In the old model, if the person who was paid to cook didn't show up for some reason, people just wouldn't eat. They wouldn't even come into the kitchen to make something. It's the same [type of] population, the same lodge, the same staff, just a different system.

Give me a cook, a custodian, and a maid, and I'll run the program by myself, by half of myself. Get me someone to do everything for everybody, and people just come and live, sleep, eat, and go to work. It's no problem, run it like a boarding house. But, if you want to run a program, have people learning something, have people using each other as a support system, then it requires a certain orientation. How do you work with people, how are you there and available without being invasive? How do you give over those responsibilities without neglecting people? How do you let the group make decisions without individuals being hurt? The more independent you want people to be, the more time you have to spend; a lot of teaching has to go on. People don't know how to run meetings, don't know how to be involved in group process, don't know how to do all the things that go into cooperatively managing a group residence.

I feel real positive about trying to do this new model, and there are signs that it is working. Ever since I came to Panta Rhei, I've been trying to push us this way, not always believing that it would work myself. When you have a house with twelve or fifteen people living in it who have long histories of being in psychiatric hospitals, all of whom have proven difficulty in managing their day-to-day lives, who suffer from quite a bit of institutionalization, being pretty passive, not doing much interacting, it's really hard to believe that these people can function as a well-oiled group. Do the practical things that need to be done. Watch out for each other. Help each other out. Get meals on the table and everything else. Particularly when you've been using a different program model that is less demanding, and it doesn't work very well.

So you go from what appears to be a less demanding model to a more demanding model and find out that the more demanding model actually works better. It's a leap of faith, but it's based on having confidence in human nature. When people have more responsibility, the more they are given, the more they take over. I've always believed that but wasn't completely sure it was going to work under these circumstances.

Before moving in, we got together as a group, both out at the hospital and here in the house. We listed all the things that needed to be done to keep the house going, and they decided how they wanted to arrange jobs and responsibilities—how long it was going to be for the term of office, who was going to ride herd on things [the coordinator]. Interestingly enough I found that the coordinator does what we do. If someone hasn't done their job, he'll just do it for them rather than get on their case. It's hard to tell people what to do even when you have that role—you're all living together.

It's exciting to see people take on responsibility on their own without constantly being told. It's exciting because it's working. If you leave people to their own devices, they'll eventually figure out among themselves the best way to carry through.

The other thing is the level of activity . . . it's catching. If you get out of bed and everyone's just sitting around not doing anything, then it sets the tone. If you get out of bed and people are talking, cooking, cleaning, whatever, it encourages you to get involved.

It seems to me people are doing much more, going places more, going places with each other. There's more of a liveliness about what goes on. And there's more activity that involves people who live there. Every week you have four people cooking, and those four people plan menus, do grocery shopping, and cook meals. That necessitates a lot of interaction with each other and with staff. Before, it might have been one client cooking and having all that interaction, and everyone else was just kind of wandering around with nothing to do. In addition there's another four people doing dishes. So that's at least eight people having pretty concentrated interaction just by virtue of what they're doing, their jobs for the day.

They designed the co-op housekeeping, and those jobs rotate once a month. But I've seen almost everyone at some time or another doing some cleaning in the kitchen. When we first get there in the morning, the kitchen is sometimes a little messy from the night before, but we don't do anything about it. Someone will show up to straighten things up. We have said the kitchen has to be paid more attention to than other parts of the house, with cooking going on. They can show up anytime to inspect, and the city is pretty regular about making their little visits. Standards have to be a little higher.

We had been working in the house for months during the remodeling and came to feel it was ours. Particularly the kitchen, where we'd plop our stuff down. We've had to make a conscious effort to not make it our territory, to put our things in the back room and hang up our coats on the hooks. I've tried to make a point to not use the phone in the kitchen. It is the heart of the house, and I don't want it to be like there's a staff person there all the time looking like they're in charge.

At first we were going to be having meetings only a few times a week, but folks were kind of drunk with freedom and weren't in the right place at the right time. We decided to have a daily meeting for a while till the routine was down. That was the arena to get people to share and talk about their experiences and get settled into the group.

One of the biggest problems people coming out of the hospital have, if they didn't already have it, is a real hard time relating to other people, having normal kinds of interactions. The more situations you can create where people are interacting with other people with practical things, with personal things, whatever, they're learning skills. Learning how to talk about how they feel, how to ask for things, how to give opinions. People take part in a communication process that's important in everything they do. Plus it is involvement and support.

I'd like to work in smaller groups. It's very difficult to sit with a group of fifteen people and give everyone a chance to talk. That size group is not conducive to personal sharing. Not conducive to people really talking about how they feel, what their concerns are. Anyone who is involved in running a group home should have some training in working with groups.

CHANGING BEDROOMS

"Having a nice room is not going to solve my problems," was one resident's comment. Nor are we suggesting such a simplistic solution, but working on the room provides an opportunity for residents to share ideas, work together, learn new skills, and build trust. Being able to control space, make choices and decisions, and change the place—even temporarily—can provide an important first step toward regaining control of one's life.

Why Bedrooms?

Just as group-home life is a secure base for reentry into the community, so the bedroom, made personal, is the base for life within the group home. Changing any space is rewarding; rewards are even greater if the space is your own.

Our view is that everyday life in a home environment is what counts; it offers a developing ground for necessary life skills. A group-home environment, by encouraging dependency, can hinder a person's efforts toward normal life, or it can support those efforts by stimulating interest, offering choices, and inviting action.

The work in this section stemmed from individual interviews and behavioral mapping of the daily use of the houses in our survey. We looked at the use of space, the activities and interactions taking place, and how people felt about their own rooms and the house in general. We wanted to know how these aspects changed as bedrooms changed and were made more supportive.

For people in group homes, the bedroom is more than just a place to sleep. It also provides residents with the sense of owning and controlling space, a place for self-expression through personalization, an arena for a range of activities—private as well as social. Yet, in most group homes, bedrooms are seriously neglected spaces.

One concern sometimes expressed by program directors and staff members is that more comfortable bedrooms will induce residents to "hole-up" there, whiling away the hours in passive solitude. Yet our research showed that with attractive private spaces of their own, just the reverse happened. Residents actually spent more time in social interaction in shared areas than previously. This is one compelling reason to choose the bedroom as the best place to start a program of change in group homes.

Another good reason is that after changes were made, time spent by residents in their rooms was more active—not more passive. Only a small increase occurred in the number of people who reported spending most of their free time in their rooms. Before changes 21 percent chose the bedroom over other rooms; after changes 29 percent did. And when asked what they used the room for besides sleeping, three out of four before changes responded "nothing." This dropped to one out of four six months after alterations were made.

While people did report using their bedrooms for a variety of activities before changes—reading, letter writing, guitar playing, being alone, listening to the radio—the range of activities after changes became more varied. More residents reported visiting with friends, reading, and listening to a radio or records than before changes, and new activities included hobbies, exercising, and drinking beer or coffee. Residents seemed to be making better use of their rooms. Obviously, simply working on and changing bedrooms for four days or four months—a one-shot deal—is not enough. Making changes to bedrooms should become a permanent added dimension to the group-home program—one for which money is regularly budgeted. The amount does not have to be great—even $500 a year can ensure that the house will feel like a growing, lively place.

Continuing enhancement and enrichment suggest that the place is never finished, that it will always require some maintenance and will need to change over time. This does not imply everlasting turmoil but recognizes that changing and adapting is a healthy part of life. It also involves a basic respect for the place and the people who use it. It recognizes the mutual, growing relationship between people and the place they live in. People have a right to a supportive environment and to influence how it will change. A growing sense of quality and caring can be encouraged and nurtured.

The involvement of people in making their

environment more supportive can be called environmental management. It is a process of incremental changes and adjustments made as a person lives or works in a place, settles in, and becomes comfortable enough to recognize and deal with inadequacies.

A particular value of making small changes over time is that it is easy to learn from small mistakes, while big ones can be discouraging (and costly). Small successes bring quick rewards. If something does not work, it is easy to put the room back the way it was or to try something else. One small improvement can be done at a time, as money becomes available. People can react to each change and offer other suggestions. A piece-by-piece approach makes it easier for real participation to occur. It leaves time for new ideas to grow.

Redesigning private spaces fosters a feeling of ownership. Before working on their rooms, most residents felt little sense of ownership. Most also saw the need for improvement, starting with better furnishings—a couch, reading lamp, mirror, or new carpeting, for example. Six months after alterations were made, 30 percent of the residents were satisfied with the changes and said that the bedroom "feels like my own." When residents were asked to describe their bedrooms, the quality of expression and description had changed. Words such as "cozy," "homelike," and "comfortable" were used. "It's my home . . . it's just me," was the way one woman put it. Residents often found something special to like about their room (14 percent before alteration, 79 percent after). The number of people citing the bedroom as their favorite room in the house increased from 36 percent to 48 percent six months after changes.

Controlling the bedroom is one part of a resident's way to regain control over life. We believe that feelings of bedroom ownership are important as a basis for feeling connected to the program and the house as a whole. A person has a stake in making it work. Without this feeling of ownership, residents see the house as merely a place to live, temporarily.

That making alterations increases a sense of ownership is also seen in the fact that residents decorated their rooms more frequently after changes; the figures jumped from 25 percent before to 60 percent six months after changes.

Even small acts of personalizing or decorating are important. Initially, only 21 percent of the residents in the survey homes had done something as simple as hang a picture or poster on the walls of their rooms; six months after completing a program of change, the figure had increased to 36 percent. Especially when moving into a new house or apartment, people just will not feel at home until their things—private possessions—are moved in and arranged. Only then do most people begin to feel comfortable. The opposite extreme is a motel room—home to no one.

Preliminaries

In most group homes, the main work force can be the actual members of the home. Painting, sanding, building, and other tasks can all be completed by the group.

Some may claim that they cannot even hammer a nail, let alone build something, but before paying a professional to do the job, even such reluctant members should be encouraged. The feeling from helping to make a place is quite special, and involvement is well worthwhile.

Skills will come with work. The trick is to get residents started. A carpenter, designer, neighbor, or tradesperson will often gladly share their skills, some for a small fee and others for the satisfaction of doing it. The Yellow Pages of the telephone book lists "Social Service Organizations" that may be helpful, or there may be retired people in the neighborhood who will volunteer their talents.

Expert advice can be valuable. Stop in at a nearby hardware store or lumberyard. Ask questions.

Some residents may benefit from adult education programs in basic carpentry, electrical wiring, and other skills. Some neighborhood and community centers also have "rehab" counseling or special do-it-yourself courses. There might even be someone living in the home who already has the skills needed.

It takes time and planning to develop the many

resources available. Three resources that we have found particularly useful in carrying out work with residents on their bedrooms are an idea center, a small workshop, and a storeroom.

The Idea Center

This should be a repository of books, magazines, catalogs, and other sources of inspiration and information. Many of the new publications specializing in home improvements, remodeling, and decorating are filled with pictures, sketches, and ideas. The house as a group might consider subscribing to a selection of these.

"How-to" pamphlets can be obtained from the hardware store, lumberyard, paint shop, or in the home-improvement section of discount stores. Manufacturers listed in the Yellow Pages will often send product information and samples of their products.

The collection of all these materials should be displayed and made available to everyone in the house. In one house, we set up an idea center in an alcove off a living room, with a small table, four chairs, a pin-up space, and some shelves (fig. 4-2). A corner of the living room, an unused pantry, a large closet, a couple of shelves in the kitchen, a hutch in the dining room would do just as well. It becomes a place for thought and discussion.

4-2. A design idea center can be used by everyone in the house.

The Workshop

While it is important that people make decisions about how to set up and use their space, it is just as important that they take a hand in the actual work. By setting up a workshop, everyone can become familiar with tools and comfortable with using them.

A simple workshop could be set up in a garage or carriage house. Local codes might not even allow a workshop in a building where people sleep, and because a workshop is usually noisy and dust generating anyway, it is best located away from the house. A workbench, hammer, C-clamps, screwdrivers, block plane, handsaw, drill, square, and some sandpaper are enough to start. With these basic tools, many projects can be accomplished: building some shelves by a desk, repairing a chair, putting up a pair of shutters, hanging a light or mirror. There are probably many of these basic tools already in the house. Others can be purchased; portable power saws, drills, sanders, and other equipment can be rented, and some neighborhoods have tool loan programs.

4-1. Books, magazines, newspapers, and catalogs can all provide design ideas.

Advice on setting up a small workshop can usually be obtained from the hardware store or lumberyard. A local carpenter who has a shop at home not only may be a source of such information but might even lend a hand.

Everyone who uses the shop must know and follow safety precautions. The necessary instructions for operating tools and working safely should be discussed as well as posted. It is essential to have good clear goggles—that must be used, especially with power saws. Equipment should be supplied with blade guards or other built-in safety devices wherever possible. The shop must be properly ventilated for sanding or painting, and face masks should be available for dust protection (fig. 4-4).

As plans develop, a sensible attitude toward fire codes and fire safety must be adopted. Codes are meant to save lives, but because they are open to interpretation, fire codes have sometimes been used inappropriately to harass group-home operators. Fire safety can be balanced with a supportive environment—without sacrificing either. Consultation with a local building inspector or fire marshall before making any changes will often head off trouble and foster good will.

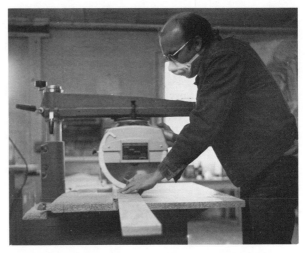

4-4. Safety rules must be learned and enforced.

The Storeroom

It is impossible for everything to be new all the time, especially in a home used by a lot of people. In institutions furnishings are usually bought at the same time and all wear out at the same time. Old and new furnishings can work together, adding character to a place and even communicating a sense of caring.

As people move in and out of a home, things naturally get moved around. Someone may need more storage and move in an extra dresser; someone else may get rid of a table to make space for exercising. Having a storeroom in the house, in the attic, basement, or garage, allows homes to keep those extra items that someone else may want in the future. The old adage tells us that one person's trash is another's treasure. When new people move into the house, they can use whatever they find in the storeroom that they may need.

A storeroom also allows people to pick up usable items—orange crates from the market, an interesting lamp found on trash day, an old ornate mirror purchased at a garage sale, free posters from a record company, simple wooden chairs from a warehouse sale, folding canvas chairs from an army-navy surplus store, and so on.

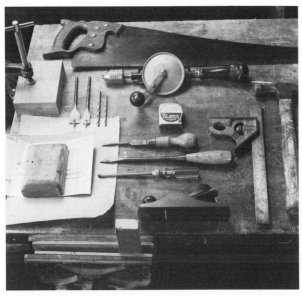

4-3. Basic tools enable residents to participate in design work.

4-5. Garages can make ideal storerooms.

4-6. A furniture storeroom allows residents to pick and choose furnishings.

As the storeroom collection grows, it also provides inspiration to people planning changes. It gets people thinking and should be arranged with care. Things can be organized and displayed so that room arrangements are suggested.

Naturally money is important; it places limits and determines the scale of the alterations. But it should not actually prevent alterations from being made. Even the smallest project can have many benefits. Doing nothing may seem economical, but it costs by depriving residents of simple privacy, growth, and change. In spite of inflation,

simple furnishings can be found for very little money, sometimes for free, with imagination. Occasionally, tradespeople will even donate materials or allow a discount, if requested.

It is surprising what can be done with even a small amount of money. One way to reduce costs is to buy second-hand furniture and refinish it. This involves removing old finishes—paint or varnish—and applying a new one.

It can be a simple job or complicated, hard work, but whatever the case, it is also very satisfying to bring a piece back to life. There are several good refinishing books available, and at least one should be consulted. Knowing a few basic rules of the game opens up the large resource of second-hand furniture to group home residents.

Making Repairs

It is encouraging when you move into a room to find it basically clean, with everything in working order. Preparing a room is like setting the stage so the new resident can move in and begin to personalize without facing the discouraging task of having to do "everything." We call this "bringing a room up to par."

It could be thought of as the first step in a process of working with a resident or as a task done in the time between one resident's moving out and another's moving in. (But it is particularly important *not* to do things ahead of time that could be part of the process of taking possession by the new residents: furniture, for instance, might best be left unchosen or unarranged.) Work can be done a bit at a time with each new resident taking part. As the room is maintained in better condition, more energy can be devoted to special things.

When a room has a basic problem, it affects the way one relates to it. One resident spent little time in his room simply because it was too cold. Eliminating the draft from the windows and repairing the radiator helped him to enjoy his room and its possibilities.

There had not been an ongoing effort of repair and maintenance in one of the houses that we worked in. Most rooms needed basic repairs and a good two-coat paint job. Small holes and cracked plaster needed to be patched. The floors all needed a good scrubbing and either painting or sanding and refinishing. Carpets and rugs needed a thorough cleaning or replacement. Windows needed washing as well as small repairs for cracks and air leaks. Lights, switch plates, and outlets needed cleaning and polishing, or replacement. Some outlets had no power because of faulty fuses.

Once major repairs are made, it is fairly simple to keep things in good order—repairing a doorknob as soon as it is loose, replacing a tile, fixing a window the day it is broken. These small efforts reflect the kind of concern that makes a place appear well cared for.

Repairs alone should not dictate the changes made in a room. Staff members or home operators may favor changes that bring a room up to par or keep it in good condition by painting and repairing. They may be concerned that personalization will not work for the next resident. Residents, on the other hand, may appreciate living in a room that is basically in order but may need something special to feel at home—a small, windowed cabinet for a shell collection, a nice rug to stretch out on, or a work surface on which to cut out dress patterns.

Which comes first, the basics or that special touch? The answer will vary with the situation. Keeping the resident at the center of the decision-making process makes it easier to decide. One resident wanted a canopy bed but decided that closet space was more important. Another decided on a drafting table that was in keeping with his interest in drawing.

As to the needs and desires of the next resident, our experience has been that a personalized room gives a positive message: if the former occupant made the room personal; so can the new occupant. It says that people in the house are encouraged to express their needs and tastes.

Getting Started

A regular house meeting or a special gathering is appropriate for introducing the idea of working on bedrooms. It is a good time to begin getting people interested and to stimulate ideas.

Not everyone will be enthusiastic right away—more people will join in as changes begin to happen in the rooms.

Consultations and Discussions

Form a working group with three or four interested residents. If there are shared bedrooms in the house, a talk with both roommates will identify problems involved in sharing the room. Then each one should be consulted individually to see what their personal needs are.

In talking with people about their rooms, use the idea center to get ideas flowing. What others have done can lend inspiration.

A trip to a department store to see the model bedrooms can be stimulating. Seeing a style, a color, or an arrangement on display might suggest solutions to residents' own problems and might also suggest uses for storeroom furnitures.

It will take some time for people to become aware of their own needs. Such awareness is a first step toward doing something about it. At the outset of our project, many people showed a low level of awareness of their own needs and the room's inadequacies. Many answered that their bedrooms were about as good as could be expected. It took some time to identify even obvious problems—sagging drapes, cracking plaster.

Nor does everyone know exactly what style is hers or his. By style, we mean a feeling or mood evoked by a room or the memories and associations it conjures. Some may like the warm, woodsy feeling of a country cabin, while others wish to create a room from *Casablanca*—all potted palms and wicker furniture. Sometimes a specific label applies—colonial or Spanish—or a single word describes it—serene or sleek.

Defining individual style is a process that may take some time and effort, but that can be a pleasure. It is a good topic for conversation in a group. Most people know something of what they like and do not like. Start with that but go beyond it. What is it that you like about this particular chair? Its lines? Its shape? What makes one room more comfortable than the other?

If it is particularly difficult for a resident to define style, he or she could try picking just one

4-7. An uncluttered, open look may appeal to some residents.

4-9. A cozy, highly personalized bedroom may fit some residents' styles.

4-8. Some residents may find a formal bedroom attractive.

thing at a time, living with it awhile before choosing another. The "style" will emerge through the process. Painting a chair, scrubbing the walls, washing the windows, starting afresh— there is energy generated just thinking about it. Start by doing one small thing. Move the bed to a different place, find a new lamp and try it out, scrub the floor, hang a woven rug on the wall or bring in some colorful pillows. This small step can be an occasion for someone to share their thoughts and to plan a next step.

On the other hand, if a resident has a strong sense of style—art deco, traditional, high tech— it should not be abandoned because of money limitations. Certain colors, patterns, or arrangements may capture the style without the expense.

Part of conceiving a style lies in seeing similarities, in finding correspondences. Furniture from three periods may all share the quality of being dark and massive or lightweight and colorful.

A person's lifestyle generally supplies hints of what is needed in the bedroom. Someone who only opens up and talks in a private situation may need a seating area for visitors. Someone who likes to lift weights might need an exercise area. A person who is withdrawn in social situations may need a private bedroom retreat. The key here—and throughout the process—is to respect styles, dreams, fantasies, and choices as much as possible within the limits of time and money.

Generally speaking, we have found most people's ideas quite practical and reasonable. If an extravagant idea presents itself, one way to deal with it is to look carefully for alternatives that have the same spirit. Try to understand the feeling that is being sought. One resident selected a deep blue for his bedroom even though this would have made it very dark. His childhood bedroom, a fondly remembered place, had been that color. We agreed that it was a nice blue but suggested using it for a graphic design rather than painting the whole room. This solution worked out very well.

Such advice must be given carefully. No one will continue to offer ideas if they feel rejected on their first try. Maybe brocade drapes are more expensive than the budget allows, but a richly colored, textured wall hanging may create a similar atmosphere.

Sometimes the room itself suggests the shape of the changes. Often there is a feature that has charm and can be the focal point—something that makes it special, personal, or different from other rooms. An alcove in a bedroom could become a cozy sleeping nook, leaving the rest of the room free; a platform with large pillows would make it a conversation area; a window seat with hanging plants makes an attractive reading place. The choice among these possibilities can be made by looking to the interests and style of the user.

Whatever the room is like, the user should take advantage of its character and explore as many options as possible. Each bedroom can be excitingly different depending on the interests, tastes, and needs of the person who lives there.

Making Plans

Sketching a few floor plans will help a resident imagine the new space and agree on changes. It is always useful to have something concrete to look at when making plans. Some residents used plans of the rooms on graph paper, positioning cutouts of beds and other furniture. Using chalk or masking tape to mark the position of each piece of furniture, right on the floor in the room itself, also helps to make planning more concrete.

It is a good idea to have a choice of several plans. The process takes time and the final decision may come only after several tentative plans have been tried out and rejected. Allowing sufficient time for decision-making means that the person involved has time to ponder and digest what each will mean.

The decision-making process by no means ends with an arrangement. If a chair in a certain place does not work, it gets moved. One roommate started painting a small wooden booth with a dark green paint. He had loved the color, but it turned out to look dull and muddy on the booth. Neither he nor his roommate liked it, so they switched to a chocolate brown. Even the choice of working on one's room or not should come from the resident. If the staff compromises residents' decision-making role, the power of the change process is weakened.

Beginning to Work

As sleeves are rolled up and work is started—gathering tools and paints, renting the sander, moving the furniture out, putting the drop cloths down—the person living in the room will need some support. It is inevitable, even if people have made the decision to change their bedrooms, that they will feel unsettled and invaded. No one really likes having their homebase disrupted, no matter how wonderful the outcome.

One way to be sure the resident's territory is respected is to agree upon a schedule that leaves the room in upheaval for as short a time as possible. People vary in their tolerance of such disruption. One resident lived blithely out of his suitcase, slept in an empty room, and never complained during the three or four days that it took to sand and varnish his floor. Another was uncomfortable with the disorder each time we worked and so we carefully reestablished order after each morning or afternoon of work. Part of the process is discovering limits, knowing when a person has had enough disruption for a while and wants to be left alone.

The more a resident participates and manages the work in the room, the less he or she will feel invaded, and the more the room will feel personal.

In one house, residents participated fully until the actual work began. Many residents were hesitant to participate in such tasks as painting and sanding. It was a common occurrence for a resident, invited to grab a paint brush and pitch in, to work for fifteen minutes and then lose interest. One resident did not want anything to do with the entire project, and we respected his wishes. At the other extreme, another resident went out, bought his own paint and materials, and did the job completely by himself. Some people happily assisted in carrying out the changes, while others were content to have it all done for them.

Participation in working on one's room might be made part of the house program. If work is expected of someone, careful, understandable limits to each task must be defined. Perhaps the whole project could be based on the condition that each person who wants to change a room make a specific work-hour commitment.

After all is said and done, the last step in this process of change is one that people all too often forget. It is very satisfying to spend time in a place that you have helped to change, to toast it with a glass of lemonade or a cold beer, to compliment the choice of fabric or furniture or the arrangement. The history of the choices made and work accomplished makes this place a special one to be in. A few photographs of such a room-warming party can commemorate the event and be enjoyed by everyone.

Room Types

Small Room

For many group-home residents, a small room can be a blessing in disguise. Occupants of large rooms sometimes have a hard time filling up the space so it does not look so empty, whereas with a small room, even limited funds and little furniture can work wonders. By small, we mean in the range of 80 to 120 square feet. A single room near the usual minimum standard of 80 square feet will require much creative effort to make it fit the range of uses it must have in a group home.

In a small space, storage is a big concern. One wall can be turned into a storage wall so that the rest of the room can be free for other things. Wall systems, combining cabinets and shelving, are neat and organized but usually expensive. Simple open shelving is much less expensive, although everything will be on display and will need dusting. Boxes or crates can be stacked against a wall or attached to it, as needed.

Built-ins, on the other hand, give a room a compact and orderly appearance. Long shelves built above the door or window frames are good solutions. They also give the appearance of a lowered ceiling in that part of the room.

Beds built on stilts—like an upper berth—leave the area below clear for other things, with the bed becoming a lowered ceiling for that special area. It could be used for a desk or couch. Use 4- by 4-inch posts and 2- by 6-inch beams for the structure, or use one of the scaffold systems made from iron pipe to form the main bed support (check the Yellow Pages or industrial supply firms). Whatever method is chosen, a safety rail for the bed should be included.

Murphy beds, platform beds, and sofa beds are all good ways to increase space when permitted under local licensure regulations. The Murphy bed folds into the wall when not in use. Sofa beds are more expensive, but their flexibility makes them economical. Platform beds can have storage underneath and can be used with extra pillows or bolsters as a couch during the day. Thin-foam, fire-resistant mattresses are inexpensive and often recommended for people who suffer from back pain. They also make comfortable seating during the day. Roll-up mattresses, Japanese futons, and other foldaway beds make a small room flexible. A bed does not have to look like or function as a bed during daytime hours.

There are other, simple ways to make a small room look and feel larger. Leaving a clear floor area in the center of the room helps and also makes circulation more convenient. Horizontal lines along floors, walls, and ceiling emphasize length, meaning that a long, low shelf, ledge, desk, table, built-in seat, or wall graphic will make a room appear longer.

Making the ceiling lower—or appearing lower—helps make the room's proportions less ver-

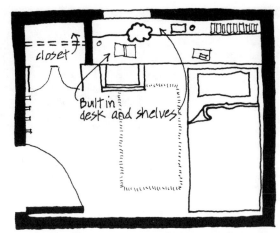

4-10. Built-in storage is a big space saver.

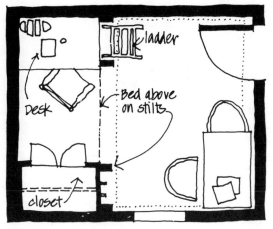

4-12. Beds can be mounted on stilts to clear floor space.

4-11. These built-in shelves provide ample out-of-the-way storage space.

4-13. A desk or work area can be added beneath a mounted bed.

tical. By appearing longer, the room will appear larger in area. To get this effect, you can paint the ceiling and a strip of the wall—say a foot and a half down—a darker warmer color than the rest of the room. Another solution is to paint all the walls and woodwork the same shade of white. The whole room will feel more expansive without the extra lines and edges.

Shared Room

In any shared bedroom—but especially in doubles—there may be both positive and negative tensions, a conflict between wanting to please and feeling resentment. Roommates must be able to get away from one another. This suggests that shared rooms need more space per person than singles do. It is sometimes unwise to wall off part of one bedroom to make two separate ones; the results might destroy the original quality of space, the resulting rooms might be awkwardly shaped, or the partitioned room might end up with inadequate heat, light, or ventilation. Sometimes, it is better to use freestanding barriers to create two spaces (fig. 4-14). Screens or free-standing wardrobes are logical items to use for partitioning. Wardrobes will also increase closet space. Grouping the furnishings of each resident is a first step, but it is inadequate as a way to obtain two spaces.

Any bedroom that is shared should be dealt with very carefully. A clear territory should be defined for each person and screened from the entry so that visitors do not have a direct view of either one. Plan circulation so that one roommate can come and go without disturbing the other (fig. 4-15). Some people want to be up and around while others are asleep.

Plan to have space for activities other than sleeping. If the roommates agree, a shared social space that leaves the two sleeping areas private can work (fig. 4-16). Individual control is most important, though, so do not exclude the possibility of having a social space in each private part of the room (fig. 4-17). If someone wants a desk or table for private visits, add that to the separate private area.

If the room has dormers or alcoves or if it is L-shaped, take advantage of the layout to help each roommate establish the identity of their part of the space.

Each roommate will need a small light so as not to disturb the other. Think of other elements that might help to define territory—different rugs for each person, shelves or ledges for possessions, and so on (fig. 4-18). It's not necessary to have each bedspread just like the next so that "no one feels cheated." Where that attitude is

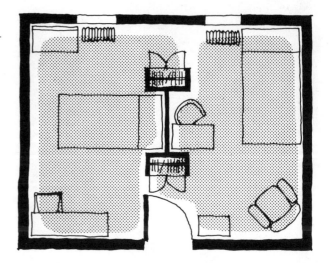

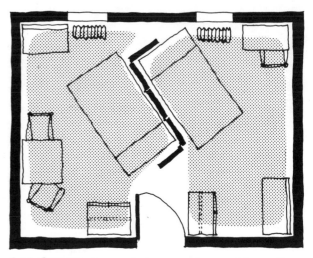

4-14. Screens or wardrobes can be used to partition shared rooms.

taken, the bedrooms end up seeming as though they belong to no one. Try to see that each person has a window or at least a direct view out. Give each person a piece of the outdoors.

Two-part Room

The widespread, long-standing image of an adult's bedroom is of a sleeping and dressing space only, unused during most of the day. This makes sense in a single family where adults have

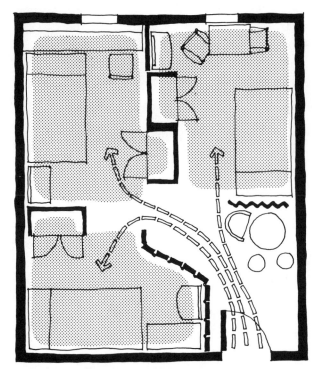

4-15. Circulation in shared rooms must be carefully planned to allow all to come and go freely.

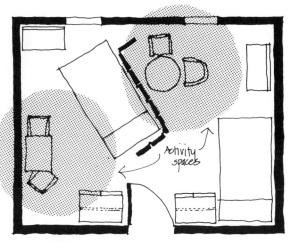

4-17. Small individual activity spaces can provide individual control in a shared room.

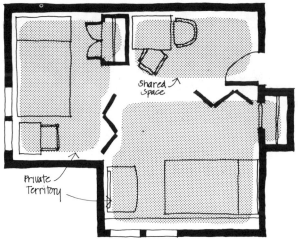

4-16. A shared social space with private sleeping areas can allow for more than sleeping.

personal control over at least one other room—kitchen, living room, den, or workshop. In a group home, each person must share these

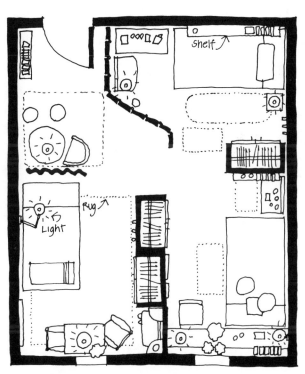

4-18. Each roommate should have furnishings that enable him or her to personalize the space.

spaces with many others who have equal control. The bedroom is the space that must be relied on to do some of what living rooms, dens, and workshops do for adults in single-family households.

As we have said, bedrooms in a group home should accommodate and encourage a variety of activities. To convey this broader definition of the space, physical cues must be provided. A new resident finding a workbench or couch in the room is given a different message than is one finding only a bed and dresser.

We have found that a multiuse message is also conveyed when a bedroom has two distinct parts: a sleeping area and a more active area (fig. 4-19). Occasionally, someone may want to set up a separate dressing space too.

For most people, to invite a casual visitor directly into a bedroom would be seen as a sexual invitation. But in a group home (or an efficiency apartment or a college dormitory), the bedroom may be the only place to visit in private. Most residents are faced with a dilemma: either entertain visitors with the bed and its sexual implications intruding or give up private visits. A day bed or sofa bed is one way of overcoming this dilemma, but such a solution may not be permitted by some licensure rules.

Even better, we think, is the two-part room. One part of the room may be public, the other private. Some people like their bed and clothes hidden from view. This can be done with a screen, curtain, or wardrobe divider. For others, the dividers do not need to be so physical. The bed and everything else may be on view, but in that case, a low bookcase, for example, can stand in the middle of the room defining two parts; or an arrangement of tables and chairs can stand on their own without a divider.

For those people who never have guests in their bedrooms, dividing rooms into two parts may still be beneficial. One area may become a work area, the other a sleeping and dressing area. One resident placed her desk in a windowless alcove to eliminate distractions and encourage concentration. In setting her space up this way, she reinforced the activity—studying. Likewise, it makes sense for someone to use a small reading chair or a workbench solely to enhance and support reading or craft work. An ordered space helps to order activities and time and encourages participation in a variety of things.

Design Elements

This section is meant to be used as a source of ideas. Residents and staff members can look through it together or separately, finding topics for discussion and solutions to specific design problems. The material is organized, roughly, from the ground up—floor treatments, walls, furnishings, display areas, and storage. It is a good idea to jot down the suggestions and ideas that most appeal to each resident.

Sanded Floors

A wood floor has a warmth and beauty that sets furniture off and gives the whole room a special character. If you have this treasure, do not paint over it or cover it up with carpet. Make the most of it. Older homes often have interesting oak or maple floors buried under layers of varnish. These floors are worth sanding and refinishing.

4-19. Separate sleeping and activity areas make a bedroom multifunctional.

4-20. Public and private sectioning of a bedroom allow residents to entertain visitors without embarrassment.

Once back in good shape, they are usually easy to maintain, for polyurethane finishes are durable and easy to clean.

For do-it-yourselfers, a drum sander can be rented, at reasonable daily rates, from equipment rental centers and hardware stores. Sanding is heavy work but worth it. Otherwise, professionals can be hired to sand and refinish; they charge by the square foot.

If the floors are in very bad shape, painting may be a better solution. A bright color such as red, yellow, or green will enliven the whole room. Because most porch and floor paints are available in a narrow range of colors, use regular enamel and put a clear urethane finish on top for durability and shine.

4-21. A beautiful wood floor can enhance any room.

Carpets and Rugs

Rugs or floor coverings have a variety of uses: they define an area, cover a bad spot on the floor, or brighten a room when hung on a wall. They add color, texture, pattern, and interest. Many people think of wall-to-wall carpeting as the ultimate in floor covering—if it can be afforded. It does have some obvious advantages, such as sound dampening, but overuse can give any place a dull sameness. To get stuck on what has become a decorating cliché is to miss some attractive alternatives, such as using area rugs or natural fiber mats to define different areas of a space. For this purpose, wool or acrylic rugs are generally a good choice. They have a good flammability rating and, in case of fire, do not produce dangerous fumes as many synthetics such as nylon or polyester do. Also, wool comes in a full range of colors. In the short run it may be expensive, but in the long run, it lasts, the colors do not fade, and the fiber stays resilient.

Whatever choice is made, look closely at the material and how it is manufactured. Generally, a tightly tufted and even surface is best for maintenance and soil resistance. Loose pile or fluffy shags are the hardest to keep clean. Nylon and some other synthetic carpets tend to produce annoying static electricity.

Throw rugs of plain cotton still exist. Braided rugs—Early American—make interesting area rugs. Be sure any throw rug lies flat and has a skidproof backing, or it will slip out from underfoot. Consider getting a pad cut to fit—they are inexpensive and make any rug or carpet stay put, last longer, and feel softer.

Woven grass or sisal floor mats are clean and add interest to any room, but sometimes they wear off the floor finish underneath. Dirt gets sifted through to the floor, so they must be taken up when cleaning or sweeping.

Choose the shape of the rug to match or contrast with some feature of the room. A round rug under a round or square table can be effective. A nice remnant of carpeting can be cut and bound by a carpet shop at a surprisingly low cost.

Platforms

Platform levels and platform beds have a crisp, modern appearance, are easy to care for, and give a room a special built-in quality. They reshape the landscape of the room by creating a change in floor level, which makes a space suggestive of a variety of uses. The platform can be used as a small stage to show off something special—bright, patterned bed sheets, a cozy reading spot, a collection of books, floor pillows that create a visiting area.

Platforms can be either movable units or built-in sections that run wall-to-wall. If need be, they can be built using plywood treated with fire retardants. Covered with carpet, they look especially neat and comfortable. A small platform underneath a favorite window, with three or four big pillows, makes a comfortable window seat.

Color

Color often sets the mood in a room. It can also affect the perception of the space. Lighter and cooler colors—blues and greens—recede and make a bedroom feel larger. More intense, warmer colors—reds and oranges especially—make a bedroom feel smaller, or make the ceiling or wall

4-22. Rugs can add color, texture, and pattern to a room.

4-23. Platform beds give a room a special built-in quality.

4-24. Platform beds near windows can double as window seats.

feel closer when used on that surface alone. Painting a small room all one color will help make it feel continuous, whole, and not as small. (See *Small Room* for other such suggestions).

The choice of color is really a matter of indi-

vidual preference but should be carefully made. Paint chips rarely represent the actual feeling that a color gives when painted on a whole wall. A better idea of the color can be gotten from a larger area and under lighting conditions similar to those in the room—probably incandescent, not fluorescent. Try taping eight or nine small chips together, or purchase a pint of the color and paint a part of the wall to make sure it is the right color for the space.

While deciding on colors, residents should be alert to colors that they like elsewhere, for example on a wall in a bank or a shop. Asking about the color and brand of paint not only is an enjoyable exchange but also means a resident may end up with a color that has been proven in action. The local paint dealer is another good source of information about colors and brands. The dealer may have some past experience with certain colors being too pale, too muddy, or too intense. At the same time, his or her taste and opinion should not override a resident's personal choice.

Window Treatments

Perhaps the most important wall elements are the windows. How they are handled may determine the room's atmosphere as much as color does. The first step is to wash the windows. In most houses, there are plenty of dingy windows, which may be giving the house a dreary atmosphere. Some houses budget money to have all the windows washed professionally on the outside once a year. It is up to occupants—residents and staff—to clean the insides. The cost in money and elbow grease is very reasonable considering the change it brings about. Everything brightens up when the windows are sparkling clean.

Many people seem unaware of the sorts of window coverings that will give visual privacy. A simple rule of thumb is that if you can see outside through the covering in daylight then probably other people can see in at night when the lights are on. To test a fabric or covering, hang it on the window, leaving the lights on in the bedroom, go outside at night, and check the window.

4-25. Plants in window spaces can diffuse light and soften a space.

4-26. Graphic wall treatments near windows can create striking effects.

Interestingly, it is almost impossible to see into a room from the outside during the day, even with the curtains wide open. But at night, with the curtains open and the lights on, the room may be on full view for the neighbors.

Hanging or placing some things, such as plants, a foot or so in from the window itself will catch and diffuse the light. This is a simple way to soften the feeling of a space. When the light strikes the plants, for example, they will glow and soften incoming light. A colorful panel of fabric, backlit by daylight, will give a burst of color. Pendants from an old crystal chandelier hung in the window, will, like prisms, bounce the colors of the rainbow off objects in the room.

Window coverings are a matter of personal taste and preference—lace curtains, heavy drapes, simple shades, wooden shutters, or bamboo roll-ups. Those who decide on curtains or drapes may not find just the right ones in catalogs or department stores and should consider making their own. The local library should have how-to books for this simple and satisfying project.

Manipulating windows and window coverings can make a room more comfortable. Keeping shades and drapes closed during the day will keep out the heat of the summer sun; opening drapes during the winter allows the sun to add warmth. For those who like having the sun in the morning and privacy at night, keep in mind the ease of opening, closing, and manipulating whatever window covering you choose.

Windows on two sides of a bedroom give good cross ventilation and counteract the harsh glare so common in rooms with only one window (fig. 4-28). Screens and storm windows should be kept in good repair. The room should take advantage of both air and the view.

Lights

Always keep the glow of a light in mind. A table lamp, a wall-mounted sconce, or even a very special fixture can give a room a clearly defined atmosphere. For example, one resident hung a fabric-covered lamp over a yellow restaurant table and found that at night ''it looked so much better and more inviting from the street than the bare bulbs and puny glass ceiling fixtures.''

Nothing kills the feel of a room more quickly than a glaring overhead light—particularly if it is

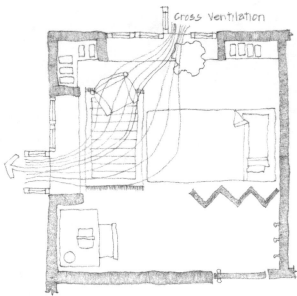

4-27. Windows on two sides of a room provide cross ventilation and counteract glare.

4-28. A variety of lighting fixtures can be used to brighten rooms.

4-29. Hanging pendant lamps provide the perfect lighting for tables.

the only source of light in the room. To create a variety of individual, special areas within a bedroom, each should be lit individually. The warm glow of a hanging pendant light is perfect over a table. Plastic or paper lanterns, a metal shade, or, wooden lamp kits (for do-it-yourselfers) can be had for very little money.

There is a big difference between a pendant lamp that directs all the light down and one that spreads light both up and down. An adjustable pendant lamp hung over a table can be brought down close to the table to create an aura of warmth and intimacy; the glow makes the area attractive. When hanging lamps, be careful to bring the cord down the wall, safely out of the way. Also be sure to have a switch right on the lamp for individual control. Lighting in other areas of the room should be as easily controlled. A bedside lamp or light switch within easy reach of the bed creates a sense of control and security. A night light might be a comfort for someone who has trouble sleeping.

In shared rooms it is important that each person have a light within arm's reach of the bed. A small high-intensity light might allow one person to read while the other sleeps.

An adjustable arm lamp puts light just where it is needed for reading, sewing, crafts, or any close work. These adjustable lighting fixtures were originally used over drafting tables, but their convenience has made them popular wherever a good spot of light is needed. They are inexpensive and available in a choice of colors and models—either wall-mounted or clamp-on.

Old-fashioned floor lamps add a touch of nostalgia and create a pool of light too. In a small room, a floor lamp may take up too much room, but in a larger space it will help to reduce the feeling of emptiness.

Mirrors

The words *miracle, marvel, admire,* and *mirror* all have *mirari,* the Latin verb meaning "to wonder at" (causing one to smile) as their root. Mirrors are a source of self-image. In institutions, where mirrors are often forbidden, patients not only lose a sense of self but are subtly told that they cannot be trusted. So, for someone who has not had one, a mirror takes on additional meaning. The presence of a mirror, with a good light and a place for grooming supplies, tells the world that the room's occupant has been encouraged to have a healthy concern for good grooming.

A full-length mirror is best for dressing. A 20-inch width and 40-inch height will be large enough for most people.

With careful positioning, mirrors can be used to reflect light from a window across the room or to extend space dramatically. Old mirrors are sometimes more interesting than new ones. An old mirror can be resilvered for a reasonable cost. An interesting old picture frame can be turned into a mirror by having a piece of mirrored glass cut to fit.

Furniture Arrangements

One of the most effective ways for a resident to make a bedroom his or her own is by rearranging the furniture according to personal preferences. Just making a tiny change—moving a dresser a few inches—can be a way of establishing territory. For those who will only spend a short time in a house, furniture arrangement is the simplest, most economical way to take control of bedroom space.

Too often, people simply leave things as they are when moving into a room that is already arranged, feeling that it does not really belong to them. One way to encourage furniture arrangement is for staff to put all the furniture in the center of the room after someone moves out and let the new resident arrange it. Another idea is to let each resident start with an empty room and choose furniture from a well-stocked storeroom. Either of these solutions could be a little intimidating to newcomers, but a few words of encour-

4-30. An adjustable arm lamp puts light right where it is wanted.

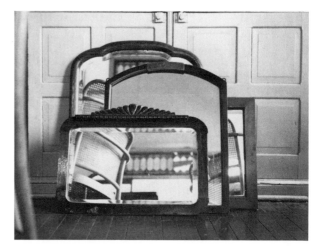

4-31. Mirrors belong in every bedroom; older mirrors can provide an interesting look.

agement from staff or an old resident, along with an offer of help, will usually get them on their way.

In many cases, furniture arrangement in bedrooms seems to follow some unwritten rules. The bed becomes the central feature, dominating the room; other furniture must then be placed flat against walls. While such a conventional layout is acceptable, breaking the rules can be liberating and refreshing.

A variety of furniture, some old and some

new, creates a friendly feeling. People have personal feelings about pieces with character: a maple rocker, an oak hutch, an intricately woven rope hammock, a wicker fanback chair, an aromatic cedar chest, simple wooden filing drawers, a hefty old desk.

Conventional furniture can be used to create some very effective layouts, although to do so more than just a bed, dresser, and nightstand—the bare minimum of bedroom elements—will be needed. Items such as freestanding shelf units, an armchair or couch, a table or desk, and wardrobe, combined with rugs and lamps, can create a very individualized space.

Figures 4-32 and 4-33 give two alternative schemes for one room using ordinary pieces of furniture. In figure 4-32, open space was the priority—for practicing the latest dance steps or spreading out dress patterns—and suggested the arrangement. A note of serenity was added with two rugs to soften the arrangements. They can be taken up when necessary.

In the second arrangement (fig. 4-33), a table and chairs right at the doorway indicated that friends visiting and workspace were most important to the occupant. Freestanding shelves or a dresser help anchor a small table in the space. A large one stands on its own more easily.

In figures 4-34 and 4-35, both arrangements use a window as a focal point around which furniture is grouped. In the layout in figure 4-34, a daybed and armchair created a conversation area. In the other, the occupant was a student and located the desk against the window wall, with a nice view (fig. 4-35). A reading chair next to it made that whole end of the room into a study.

If the room has a dormer, alcove, or other indentation in its walls, as in figures 4-36 and 4-37, it can become a special defined area. Putting the bed there leaves the rest of the room open for other things. Or it can become a sitting area. Other solutions include setting up a work area or a special place for clothes.

If the bed is to be the featured element, put it right in the center of the room. Place shelves or a dresser or a chest against the headboard (fig. 4-38). There still may be enough room to place a narrow desk or table with chairs along one wall. Each resident should try two or three different arrangements before settling in, and might team

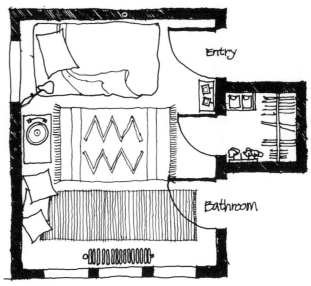

10 X 12

4-32. This layout provides plenty of open space; rugs add warmth.

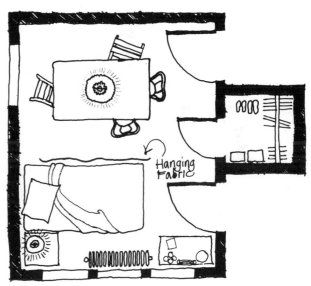

10 X 12

4-33. This furniture arrangement, with table and chairs prominent, gives social space priority.

up with the person next door to try different arrangements in each of the rooms.

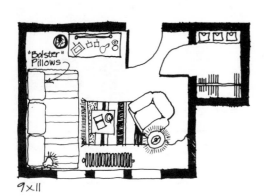

9×11

4-34. This layout creates a conversation space beside the windows.

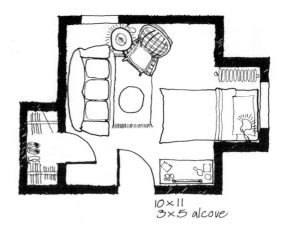

10×11
3×5 alcove

4-36. Placing a bed in a dormer space leaves the rest of the room open for other uses.

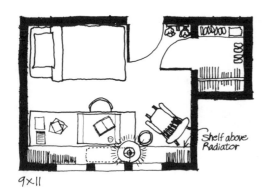

Shelf above Radiator

9×11

4-35. This arrangement makes the window space into a study area.

Entryways—Special Handling

An entry space in a private bedroom gives the occupant the ability to exercise more control over the space. It makes the room feel just a bit more private, and it gives an added dimension because the room has parts rather than being one whole.

By shape alone some rooms have a natural entry space (fig. 4-39). In other rooms, there are several ways to create one (fig. 4-40).

Freestanding screens can be used to shield a

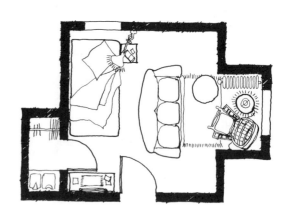

4-37. A dormer can also be used as a sitting area.

bedroom from unwanted visual intrusions. A brightly colored fabric, hanging 3 or 4 feet in from the door, can work too.

Furniture can create an entry. A wardrobe, high dresser, or bookshelves placed near the doorway forms a sort of foyer or vestibule.

In a shared bedroom, freestanding wardrobes or dressers will create an entry space and define each person's part of the room. If the back of a

4-38. In this bedroom, the bed is the featured element.

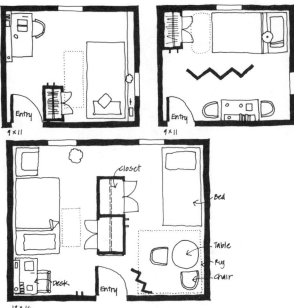

4-40. Screens and wardrobes can be used to create entry spaces.

4-39. An entry space to a bedroom gives the resident more control over the room.

dresser is unsightly, it can be covered with fabric, painted a bright color, or turned into a bulletin board by tacking cork onto it.

Screens

Screens can give visual privacy. They can be bought or built, are generally lightweight, and are easy to move around. Screens can also be improvised by hanging an attractive fabric from the ceiling, with the help of a dowel and ceiling hooks. Louvered shutters, which come in a variety of sizes and shapes, can be hinged together to create a screen of almost any size, which can be opened out or left partly closed (fig. 4-42). Unfinished shutters are inexpensive and can be painted or varnished.

Interesting old doors, sometimes found in attics or basements, hinged together will make a screen. Building salvage and wrecking companies usually sell old doors, often with special moldings or panels that can be painted bright colors or refinished. We recycled an old oak office par-

4-41. Screens provide privacy as well as decoration.

4-42. Louvered shutters make effective screens.

tition that had wooden panels topped with frosted glass. It was heavier than most screens but made a nice solid divider in a large bedroom.

Sheets of plywood cut into different shapes and sizes create unusual screens. Edges can be curved and shaped into patterns; perforations can be made.

Not all screens must be solid. A latticework screen, made with lattice obtained from a lum-

beryard, can be attached to a frame made with 2×2 or 2×4 lumber in a vertical, horizontal, or diagonal pattern. This creates a barrier while still letting light through. It can be painted white or left natural. A beaded curtain hung next to a reading desk creates an airy effect. A large bamboo roll-up shade hung across the middle of a room becomes a quick flexible divider. Fabric stretched over frames made from 2×2 lumber is another possibility—staple the fabric to the wood and hinge the sections together.

Tables and Chairs

Tables make a functional addition to a bedroom. They can be used comfortably as desks or workspaces. With a few chairs, a "social space" is suggested. Having a table between you and your guest also eases the tension of entertaining in the bedroom.

Tabletops and bases can be bought separately from restaurant equipment stores. If secondhand, they may need some work, but they are a bargain just the same. The pedestal bases usually sold with these tops are appropriate for small tables, leaving the space underneath clear for chairs.

Electric cable spools make interesting tables, and the local power or phone company might actually give them away. A hollow-core door (inexpensive at lumberyards) set on sawhorses or crates is another variation and makes an inexpensive desk (fig. 4-45). One house found an old wooden booth—a table with two high-backed benches—at a rummage sale. When it was sanded, painted, and placed in an alcove, it became an inviting place to gather (fig. 4-46).

Tables need to be carefully located in a room. Placed against a wall or a screen, the table feels connected to the space and secure. In a corner, it becomes a cozy place. A table placed next to a bed becomes a nightstand as well as a place to study. Located by a window, it makes room for plants and lets the user take advantage of good reading light.

Experiment by moving a table around in the room—place it lengthwise against a wall or protruding from a wall or in the middle of the room.

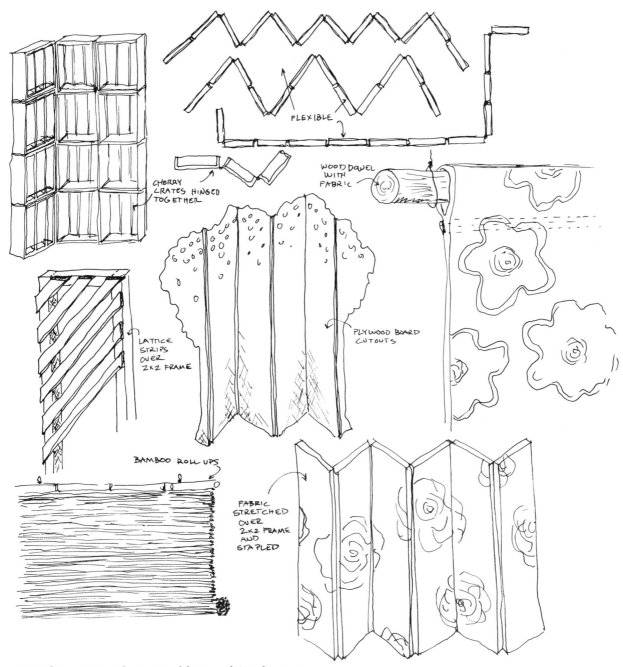

4-43. Screens can be created in a variety of ways.

Remember that in restaurants, people invariably sit in corners or against a wall if they can.

As for chairs, inexpensive, wooden or metal folding chairs are widely available. They are easily stored and brought out when needed. One storage solution is to hang them on big bicycle

4-44. A table and chair create an inviting work space.

4-46. Wooden booths make appealing social spaces.

4-45. A hollow-core door can be mounted and used as a desk.

hooks on the wall. Foam pads or cushions will make the seats a little softer and brighten them up. Light bentwood chairs are durable and easy to care for. Besides being attractive, they blend in well with a variety of different styles of furniture.

Canopies

Most people associate canopies with beds, but they can be hung over a special part of the room or over the whole place to lower the ceiling. Canopies can be either quick and informal or a permanent architectural change. They will in any case make the room distinctive and appealing, cozy and sheltered.

Hanging methods vary according to need. Lightweight fabric can be draped over wires stretched tautly from hooks in your walls (fig. 4-47). With loops sewn onto the edges of the fabric, it can be hung from hooks (fig. 4-48).

For safety's sake, material should be treated for fire resistance. Trevira polyester, some acrylic materials, and wool are fire-resistant; other fabrics are pretreated, although they may lose their resistance after being washed a few times.

The weight of the material is another factor to

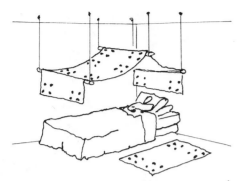

4-47. Canopies can be hung from wires and dowels.

are extremely light and graceful. They can be draped to resemble the inside of a tent. A colored parachute is very bright, particularly if there is light coming from above (fig. 4-50). This sort of canopy works as well over an activity area as over a bed, where it creates a cozy sleeping nest.

A lowered section of ceiling running from wall to wall—a soffit—over a bed, table, or entry will define a space in a more permanent way. It should be kept simple and made from the same materials as the original ceiling so that it blends in. Soffits can also be used to "box in" and hide ugly pipes or damaged ceilings. This is not a simple project, however, and may require the help of a builder.

4-48. Loops in the canopy allow it to be hung from hooks in the walls.

be considered. Felt may be too stiff for the shape of the canopy, while summerweight wool is light and hangs gracefully (fig. 4-49). Army surplus stores often carry used parachutes, which

4-49. Summerweight wool is a light fabric that drapes gracefully.

4-51. Graphics add color and interest to a room.

4-50. A parachute canopy can give a dramatic look to a room.

Graphics

Graphics work best when they are the personal expression of an individual. Too often, though, graphics are thought of as a "quick cure" to brighten up the place. They are done in an institutional way, making each room similar and giving each resident little control over what is done. They end up not belonging to the people who use the space but rather to the volunteers' association that got the art students to paint them. Sometimes this leads more to resentment than appreciation.

However, graphics can be used to bring bold bits of color into a bedroom. Rather than having every room painted the same, one room may have a bird, another a sunburst (fig. 4-51), and a third a stenciled flower border. Instead of having every bedroom door boldly painted with its room number, one might have the occupant's giant initial on it (fig. 4-52), another, diagonal stripes in favorite colors. There are numerous ways to bring color into the rooms in graphic ways.

Some residents may want to do the work themselves; others may want to decide what is done but let others do the actual painting. Either

4-52. Residents can choose graphics to identify and personalize their rooms.

way, it remains a matter of personal choice. One resident painted clouds on the ceiling; another painted a false window with a view of the Eiffel Tower.

A graphic shape can highlight a special area in the room: a half circle by a stereo center (fig. 4-53), a stripe around a shelf that displays favorite things, or a circle around the window. Low horizontal stripes can make the room seem longer and make the scale more comfortable by accenting the lower half of the space. Graphics can add a large area of color without making a room seem too dark (fig. 4-54).

In the beginning, the graphic should be simple—stripes and bands of color, which are easy to paint. Two-inch angled edging brushes will help keep edges crisp and clean, once the lines are sketched in pencil. There are also supergraphic kits available that have stencils and other hints to help in laying out graphic ideas.

Graphics may not solve basic bedroom problems, but for some people they are very important. One person may not even think about them until the storage problems are solved, while another may want to paint a sunset on the wall right away.

Pictures, posters, snapshots, and so on can be hung without ruining the walls by using good-quality, aluminum-headed push pins. Pushed directly into the plaster, they will not do much damage and are fine for such things as posters, fabric hangings, matted prints, a woven handbag or Chinese paper dragon. However, they are inappropriate for heavier things—a lithograph, framed and glassed, for example. Then, a more elaborate approach is required. Using the traditional picture molding will not cause wall damage. Picture moldings are available either as a wooden trim, sold at lumberyards, or as aluminum strips, sold at home improvement centers. They are nailed or screwed to the wall, usually near the ceiling. Metal clips or S-hooks attach to the molding, and wires or translucent fishing line attach, in turn, to the picture below (fig. 4-55).

Somewhere between the simplicity of push pins and the permanence of picture moldings are various other types of picture hangers—nailed in or stuck on. They do limited damage and, properly sized, are dependable. A "clip strip" can be made by attaching spring-action clothespins or

4-53. Special spaces can be highlighted with graphics.

4-54. Graphics can add bright color without overwhelming the space.

metal clips to a lath strip (sold at lumberyards) as in figure 4-56. Screw the strip to the wall at sitting or standing level, or put up a whole field of them (fig. 4-57). An even simpler version is a cord hung along the wall with plenty of spring-action clothespins clipped to it. From this can be hung postcards, wish lists, cartoons, and calendars.

Tackboards are another solution. Their presence encourages personalization by each new resident. Cork works well and comes in squares that can be glued directly to the wall. Homasote, made from pressed paper, is another inexpensive tackboard material. Two or three sheets, covered with fabric and nailed side by side (with small

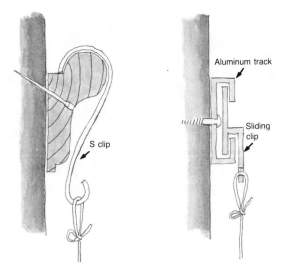

4-55. S-hooks or metal clips, with wire attached, can be used to hang pictures.

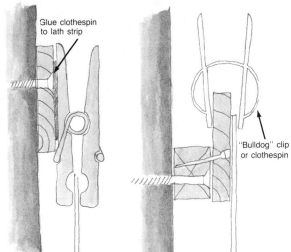

4-56. Clothespins or metal clips attached to lath strips make handy clip strips.

4-57. Rows of clip strips make useful and attractive display areas.

Display Ledges

While tacking directly to walls or tack boards is one solution to display, it sometimes results in wall damage, which, though minimal, may be discouraging for the next inhabitant of a room. Instead, by using ledges effectively, people can find ample space to hang, place, lean, and pin things to their heart's content without damaging the walls. The character of a ledge in a room changes from occupant to occupant. One woman had her ledge covered with lace doilies and lined with pots of African violets. After she left, a man moved in and propped a cork board on the ledge where he pinned photos of his show dog, Lulu.

If two ledges, top and bottom, are used together, Homasote or cork-covered panels can slide into place between them for a pin-up surface; or a piece of blackboard can be used instead for writing reminders. Either of the ledges can be put up and used alone—the upper with pegs and

finishing nails), make a tack wall that is sound-absorbent and insulates as well.

For those who are unsure about framing, a do-it-yourself frame shop will do the cutting and advise on assembly. Metal frame kits from art-supply or department stores are also easy to assemble.

4-58. Ledges can serve as effective display spaces.

hooks for extra hanging space or the lower, which is generally wider, as an objects space. A short ledge might take the place of a nightstand, with room for a radio, alarm clock, glasses, book, and a water glass. A ledge with a mirror can be used as a vanity with a place for cosmetics or toiletries.

Placed carefully in a room, ledges will contribute to its appearance. Handsome ledges can be made from part of an old mantlepiece, a piece of wainscot paneling, chair moldings, or plate rails.

4-59. Ledges can provide spaces for clothing storage as well as for display.

4-60. Ledges can be created in a variety of interesting shapes.

4-62. Ledges can be used to hold ornamental items.

4-61. A ledge with a mirror could serve as a vanity space.

Shelves

One of the easiest and most flexible ways to set up shelving is with bricks and boards and stock lumber (fig. 4-65).

Another inexpensive solution is to use inexpensive metal L-brackets sold in a variety of sizes in hardware stores. Two or three, screwed to the wall and painted the same color, will support each shelf. However, such a shelf may not be strong enough for very heavy things.

Another common system consists of standards—metal strips that are screwed vertically

4-63. Even puppets and eggs can find space hanging from ledges.

into the wall—into which shelf-support brackets are inserted. Most hardware stores sell standards in precut lengths, but they can then be cut to any length with a hacksaw. The brackets come in

pensive system can be used to put up one shelf or a whole storage wall. An alternative, metal industrial shelving, is popular and is now sold in small units at department and furniture stores. The original units can be found at office supply or industrial hardware stores.

Fruit crates from the local farmer's market can be stacked and used as shelves or supports for wood shelves. The labels on the crates add a decorative touch and some color, or they can be painted a bright color. A wide, low shelf could double as a seat. A long, narrow piece of plywood supported by stock lumber resting on the floor might work very nicely with a few pillows.

Not all shelves need to be long. A short little shelf can hold one favorite object and be put up in a special place in the room. Narrow shelves are best for special items—china frogs, favorite pipes, model cars, and pictures.

In putting up a shelf, the visual effect must be considered. Does it tie in with other lines in the room and fit into the space? Shelves running the entire length of a short stretch of wall can be very effective. In a nook or alcove the shelves might also run the entire length or from floor to ceiling to emphasize the height of the space. On the other hand, keeping them low establishes a comfortable scale.

4-64. Display ledges can be made from moldings or plate rails.

4-65. Boards propped on bricks make flexible, inexpensive shelving.

a range of sizes for boards of different widths. The shelves can be moved up and down easily to fit books, records, or other objects. This inex-

4-66. Narrow, as well as wide, shelving can be used for display.

4-67. Shelves are not just for books.

4-68. Metal milk crates provide handy box storage.

Box Storage

"I need a nightstand by my bed," said one resident, "and some more storage—but this room is too small to fit them in." In this case, boxes are one answer. Fruit crates with shelves nailed in, metal milk crates, or "off the shelf" commercially produced storage cubes can be attached directly to the wall.

However, if stacked and left freestanding, they are very versatile. They can be arranged and rearranged by each new resident. Lined up, they create a long, low shelf, next to the bed, for example. Stacked, they double as a bureau.

Boxes can be used to define space and add scale and detail to a room. One resident used them to enclose an area around her table. They can be arranged to define boundaries between different areas of a shared bedroom, providing storage in both directions as well.

Some furniture and design stores carry box systems. Each has a range of accessories, including doors, drawers, or record and wine racks. They are expensive but certainly versatile. Pala-set and Cubex are plastic cubes just over a foot long on each side; Cado is a set of boards that can be assembled into cubes or shelf boxes.

Home improvement centers have kits to make particle-board cubes, a foot on each side. Such

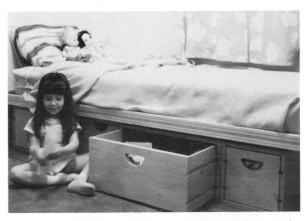

4-69. Box storage systems can be placed in out-of-the-way spots.

cubes can also be made with particle board purchased and cut at a lumberyard. They should be about $13\frac{1}{2}$ inches on the inside, big enough to hold records, files, wine bottles, and large books.

Portable Storage

Moving is certainly part of group-home life. "I'm packed and ready to go," said residents to us over and over. Some actually did leave the next day, while others stayed on another few months. Inexpensive storage that can be taken along when the time comes to move is very useful. It makes the move easier and adds the comfort of a familiar object to the new surroundings. It is easy enough to pick up an old (or new, for that matter) trunk, chest, suitcase, or footlocker. Lined with bright paper and with the lid propped open, an old trunk can add a romantic flair to a room as well as be a practical place to keep many things.

Hooks and Pegs

Hooks and pegs easily increase storage space. They can be added to ledges, lined up in a row, for extra clothes storage. In a zigzag pattern, they are decorative as well. They can be hung in two rows, one above the other, or attached to a backing board, screwed to the wall. Individual hooks can be for special items—or a row of them across a whole wall will hold everything in sight. Hooks and pegs put things right at hand and make a room—even for someone with few possessions—seem full.

An accordion hatrack (fig. 4-71) is inexpensive and easy to find. A coat tree makes another useful addition to any bedroom.

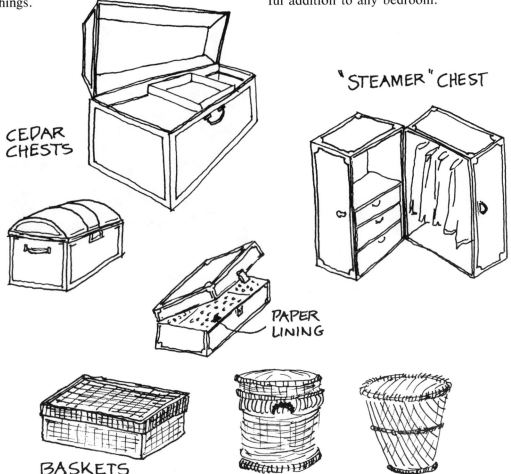

CEDAR CHESTS

"STEAMER" CHEST

PAPER LINING

BASKETS

4-70. Chests, baskets, and trunks allow for portable storage.

4-71. An accordion hatrack is an inexpensive, decorative storage item.

Closets

Because closet space is usually at a premium, it should be most carefully considered.

A newly purchased wardrobe is liable to be quite expensive, but second-hand ones can often be found at auctions or thrift or antique shops. Old-fashioned armoires will probably be large and ornate and very expensive. However, other pieces can be modified into closet space. An old hutch with the shelves removed could work, with a bar for hangers and some hooks added. A little fresh paint outside and a fabric covering inside will make it most attractive.

Or a wardrobe can be made from scratch by building a plywood shell and adding louvered doors or a macramé curtain or other covering.

Gym lockers or industrial cabinets add a high tech look. Even institutional metal furniture can take on a whole new look when painted in bright colors. Other solutions include a freestanding coat rack with hangers, a rod suspended from the ceiling as a hanger bar, or a wall-mounted frame for a clothes bar. With any of these, a dressing corner can be sectioned off with a screen.

With the existing closet, many problems are often caused simply by poor lighting in the depths. A fluorescent lamp complete with plug-in cord and switch is easy to come by and will make a big difference.

4-72. Inexpensive closets can be made with plywood and louvered doors.

To increase shelf, drawer, or hanger space, there are many plastic or metal systems available at discount or department stores. Adding more shelves in the same space will mean that the stacks of clothes are shorter and easier to pick through. If a closet or wardrobe is to be shared, different areas must be clearly divided and labeled.

Visible storage—hooks and pegs, open shelves—can help decorate a room that would otherwise look barren. This is especially important for someone with only a few possessions.

Bedroom Stories

The following three stories are drawn from bedroom remodeling projects conducted at the Panta Rhei Lodge. The lodge program was operated for adult men and women with a history of repeated admissions or long-term hospitalization in state mental institutions. The program was open to those aged eighteen and over, with the only stipulation being that residents be capable of working in the lodge-operated business, an office-cleaning service. The bedroom remodeling project lasted two and one-half months, during which time a member of the design group worked individually with each resident to make his or her bedroom personally supportive.

BEDROOM STORY ONE:

BILL

Bill was interested in setting a mood and creating a dramatic setting in his bedroom. He wanted a "place to come home from work, lie down with a beer, be able to reach over and turn on the TV or stereo. A good place where I could entertain, too."

He was attracted to sleek, modern materials, as opposed to traditional ones. "Smooth, not woody," he said. He liked modern lines and easy-to-maintain designs, which meant there was not too much to clean around. When he looked through designer books, he was attracted to dark colors, to simple restrained forms and layouts. The furniture he liked was painted or plastic with chrome, in dark or intense colors. Over the weeks of working together, Bill's vision of his ideal room began to change along with his perception of his own needs.

Bill decided to use a screen to separate the sleeping area from the activity area where he could entertain a friend. After the screen was in and the private and public spaces were defined, Bill lived with the arrangement for a few days and then changed it, moving the table and chairs to the other side of the screen, with the bed. This created a small entry space that acted as a buffer between the private part of the room and the doorway.

Throughout the work on the room, Bill seemed to have highs and lows. He was sometimes enthusiastic about the changes, sometimes aloof from them. He also worried at times that too much attention and money were being devoted to his room. Occasionally, his original flash of design inspiration faded, and he became less sure about the themes that he had formerly been so clear about. Some of the original feeling did come through in the finished room though. Inside, behind the screen, it felt protected, private, exclusive—a place where Bill could do what he wanted. The soft lighting, screen, and stereo next to the bed helped set the right stage.

As we went along, a gap became apparent between Bill's imagined room and his lifestyle. The image he had was of a pristine space, yet

4-73. Bill originally used a screen to define public and private space.

Bill himself had great difficulty keeping his room tidy and clean. Adding pegs on the ledge board as an easier place to hang clothes was an attempt to handle this problem. Even after the changes, though, Bill often piled his clothes on one of the chairs that went with the table.

Bill did show signs of coming to terms with his lifestyle. He changed his mind about buying a comforter because his dog, Peppy, was far too inclined to chew on things. It was either "the whole room to be admired" or the dog. Bill chose to keep the dog and adjust his ideal.

It is hard to convey Bill's complexity and the complexity of working with him on his room. In later describing the process of change, he was in part critical, which helped us to push the boundaries of our understanding a bit. He also learned about self-help from the experience, reaffirming other parts of our attempt.

If there was a sense of disappointment for us, it was that we had not heard the criticisms and affirmations from Bill during the process. If we could go back, we might encourage him to explore his feelings—and us, ours—on the spot. The following are his comments after the remodeling process was finished.

"At first when I moved to Panta Rhei, I wasn't really aware of the full program and how it was run. I thought it was going to be until I

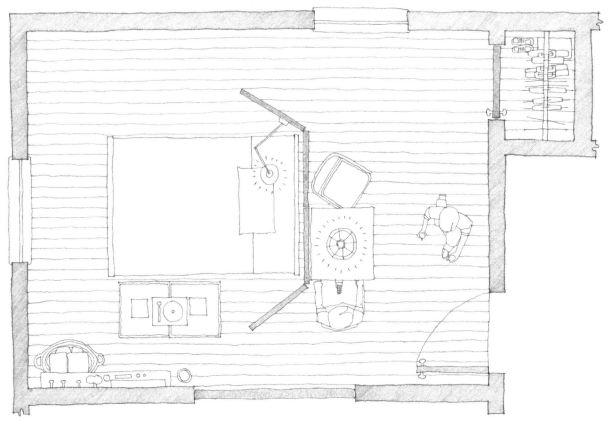

4-74. Bill's first layout.

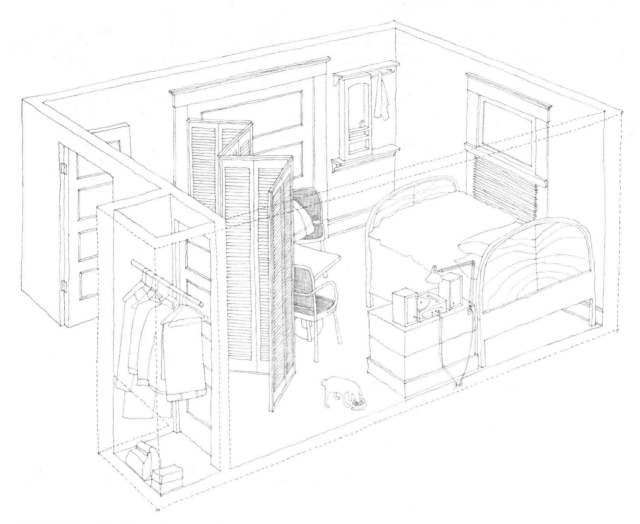

4-75. Bill's revised layout.

felt as though I could do better elsewhere or something like that, you know.

"I was kind of frustrated because I couldn't understand why it [the redecorating] should be all upon me. I mean, here I was in the hospital for two years—and I don't know how many times before that, in and out—and I have to come out and make decisions on how the room should be. I'm not going to be in the room very long, and other people are going to be moving in from the hospital. It wasn't home; it wasn't mine. I was paying rent for it, but it wasn't mine. I had no room, no home, no furniture, no clothes. I felt very confused and mixed up and very angry—that's how it usually comes out to be with me, a lot of anger.

"While working, I began to feel a little bit of relief. I felt that Panta Rhei was doing something for me, but it was so instant. In the hospital everyone does things for you. You cut your finger, and they put a Band-aid on it for you. And still I'm in a sheltered situation. And I keep thinking about how, before I had this nervous breakdown, how I used to live. I can't touch myself now; I mean, I had two jobs, a brand new car, friends. I have nothing now.

"And all I could find out, when I got involved in something deep—like when we were refinishing that screen in my room—I thought that was pretty heavy because I'd never done it before—and I found out that you could take an old piece of wood and make it look new just by using some liquid and a rag. Make it look different, make it look antique or modern, and I thought that was pretty hip.

"So I felt like if I could just stick with this program, 'cause it's been keeping me out of the hospital, maybe with what I think and what I know, maybe we could work something out that could benefit other people like me.

"I didn't like the idea of getting off work and coming home and doing more work. Because—well—I've always thought that work was something you have to do—and it should be made as interesting as possible so it wouldn't be so confounded hard. You know, you dread it in the morning, to get up and go to it, and then when you get the hang of it, you don't want to leave it. It just gets to be a big pressure.

"So I worked with [all of] you, and I saw how you made things easier when we worked together.

"I think I did a little bit on my own. I left something there that was really me—like the screen—the wooden screen. I really thought of that. After looking at some designer magazines and going through them, I picked that out myself. So it was really me.

"I didn't use the room very much, because I couldn't smoke in it. And I had very few friends there as company. It was so intimate there: I didn't know anybody that way. But sometimes Steve would come by to work or do something, and I found it very relaxing just being there with him. Looking around at the new stuff and seeing the old stuff sort of rebuilt. That made me feel good.

"And I've never liked old furniture. I always thought that stuff should be cast away, like for poor people—or just throw it away. I never did see any sense in varnishing and shellacking and going over something like that when I'd rather just go pay the money for expensive stuff that was really nice. To keep forever. But I learned how to go to the Goodwill and pick out something like a bedroom suite, because you're not going to have company in your bedroom; and buy a brand new living room set, since you're going to have company coming there. I see that you don't have to go just broke like I always tended to get. If I'd never met you people, I'd have probably always been broke, if I ever came in contact with any money.

"Working is important. My father worked thirty-five years for New York Central railroad, and I remember how things should have been, and they weren't. And how I used to get beatings from him because I didn't do my job at home. But to me I think it is vital, it's like exercise, it's like play, you know. It's fun. It's healthy for you. And you can accomplish an image out of work.

"I didn't like the idea of having to move out. I'm thirty-six years old, and I want to stop and settle down someplace. I don't care where it's at. I'm just tired of moving from one apartment to the next, one room to the next, one space to the next, taking a bag here—I'm fed up with it. If I leave again, I want to leave everything and say, 'You can have it. Goodbye, I'm free.'

"Of course I would like to live someplace and own something—like furniture, a car, things of that nature. And working with people around here, I saw where I didn't have to spend a lot of money. Of course, if I had money, I saw where I could spend it too.

"So I've decided to save some money and buy a trailer home and a small car and move around and live kind of close. I think that's about all I want to say. I have said a lot."

BEDROOM STORY TWO:
ANDREW

Andrew was a team player, a good steady worker, friendly, agreeable— always neat and orderly. But getting opinions, suggestions, or requests out of him was like pulling teeth. Maybe his passive and polite qualities made it hard for him to state his needs and make decisions. His attitude was always that "everything was okay," or "they" probably would not allow changes anyway.

Making decisions with Andrew worked best as a series of small steps. We capitalized on every glimmer of interest or energy from him.

Seeing many ideas, one after another, in a decorator book seemed confusing to Andrew. It was better to sit and talk about just a few ideas in one book or magazine. Once he discovered and expressed some general likes and dislikes, a few plans were drawn up for him to choose from. Limiting the number of alternatives or even eliminating them altogether seemed to make the process easier.

Because Andrew did not make many demands or requests, we looked to the room for suggestions as well. An alcove, formed by a dormer window, was the feature that influenced the three plans that were drawn up. Andrew chose the one with a booth in the alcove because he felt he could make use of it.

It was hard for him to choose a color for the booth. He would choose a range of colors but not the color itself. He seemed to add complications: "Well, the floor is red now, so red would look nice in the booth, but in the future they may paint the floor another color and then red wouldn't be good." In the end, we had to make the decision for him.

Andrew, like many other residents, found it hard to recognize desires and express them. He was a challenging person to work with,

4-76. Andrew liked the idea of a booth in the alcove.

one with whom only subtle changes were apparent during the initial process. Any small signs were important. At one point when talking about the color to use on the walls of his room, Andrew was drawn to a memory. He had once painted a room ivory—"that would be a good color." The sharing of a memory was honored and built upon. We were not always quick enough to do that with him, but we learned as we went along.

Only in reviewing the whole period were we aware of how Andrew had changed. His clothes looked more carefully chosen, and new ones had been added to his wardrobe. There were also changes in his behavior and ways of interacting with other people. He was able to befriend another resident who had been in the army as Andrew had. He and his friend used Andrew's room as their favorite place to meet, drink a beer, and talk.

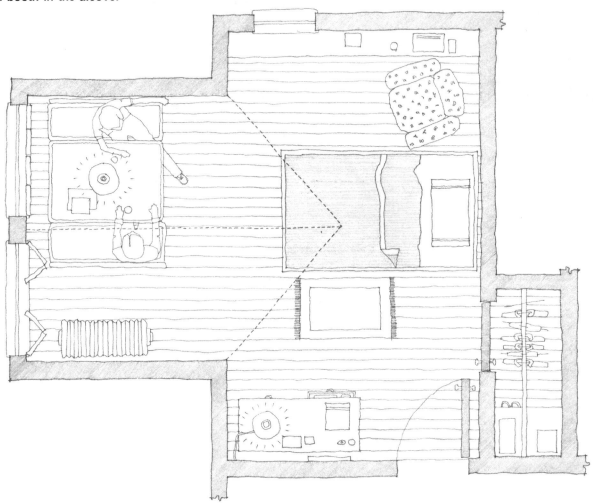

4-77. One of three plans shown to Andrew—and the one he chose.

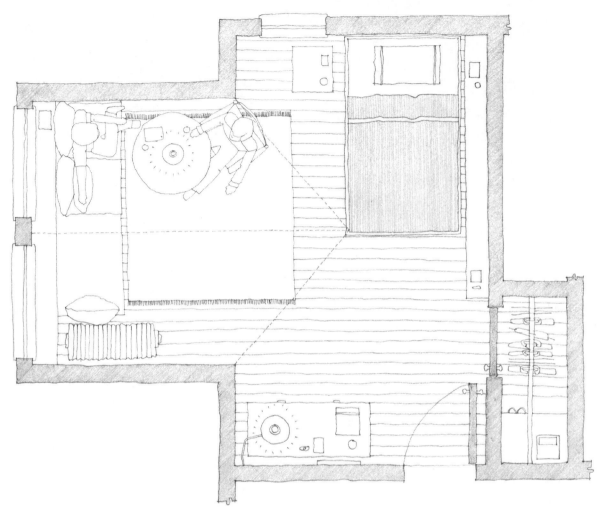

4-78. Another layout suggested to Andrew.

Andrew: As far as my room is concerned, I think it was made a lot nicer. Putting the table and chairs in there. One drawback is the cleaning part of it. It's delicate to clean.
It's hard to clean?

Andrew: It's delicate to clean. But the booth and chairs are the best part. I like it all right!
Does that make a difference in the way you feel about your room?

Andrew: Yeah, I think it does. I spend more time there. It's made it a lot nicer.
Did you learn anything from the project that you might take with you?

Andrew: That long shelf he put in there, that was all right.
Do you think you might do that in another place? How do you find it useful?

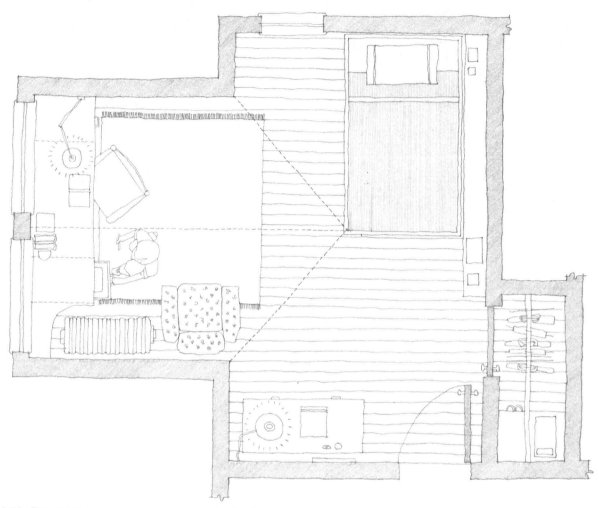

4-79. The third arrangement proposed to Andrew.

Andrew: Sometimes I just put things on it. If you sit down in the chair, you've
           got a shelf to set things on . . . a glass or a cup or something.
               Would you like to change anything about the room?
Andrew: A little wall-to-wall carpeting, maybe.
               Have you yourself done anything to the room?
Andrew: No.
               Would you?
Andrew: Not that I could think of.
               Have you moved anything around in your room?
Andrew: No, not at all.
               But you know you can if you want?
Andrew: Yeah. It's nice the way it is.

BEDROOM STORY THREE:
CAROL

Two days after Carol moved into the house, the only other woman resident moved out. Carol felt ill at ease then. The men in the house were all strangers, and she did not know if she was safe. She thought they might have designs on her, and she was more than half right. The first few weeks she spent sleeping late every morning, coming down only for lunch and supper, going out to work, and coming back to lock herself in her room in the evening. She had a television there, and she felt safe.

As she got to know the men on her work crew, she began to trust them and felt more confident in the house with them. A kind of friendship developed as she shopped for snacks to share after work or sewed on a button for someone. She began to talk about feeling good in the house. "The guys here are great," she said more than once.

Her contact with the larger community was similar. During work on her bedroom, she was asked to go shopping to look at curtains and bedspreads for her room. The day was set, it sounded like fun, but when the time came, she could not do it. Crowds frightened her, and the plan was just too much for her to handle. She did help do the food shopping for the house and slowly became acquainted with the neighborhood—banking at the nearby bank, stopping to buy a few things at Woolworth's. As time went on, she was moving in broader and broader circles out from the house.

When Carol started talking about how to change her room, it became apparent that she was quite concerned with her daily schedule and the order in her life in general. She had a big calendar propped up on the mantlepiece and kept daily lists of everything she was to do on a given day.

Her need for order remained a theme throughout alterations. Her request for more hang-up space was related to it. At one point, she was enthusiastic about having her floor sanded, but when she saw another resident going through the process—furniture in the hall, bedroom inaccessible for three or four days—she quickly changed her mind. It was too much disruption for her.

Carol's room itself was quite large by regulation square-footage standards. Figuring 60 square feet per person, this room could have had four people crammed into it, yet it did not seem overly large for Carol alone. The room was rather traditional, almost old-fashioned, as might be expected in a large old house. The woodwork, moldings, and trim around doors and windows, and even the gas fireplace, all made their contribution to the charm of the place.

When Carol first moved in she had rearranged the furniture so that she could watch television from the bed, which was in the middle of the room. The other furnishings—two dressers and two upholstered chairs—were distributed around the four walls. She was not particularly attached to or comfortable with this arrangement and was looking for suggestions and advice.

Even so, it very soon became clear that the room would not be undergoing major transformations for that would be too disruptive for Carol. While we were experimenting with different rearrangements, we decided to paint. The room had been wallpapered long ago and a few spots were peeling. One corner had been damaged from a leak in the roof. After some patching and spackling, the room was ready to paint. Unlike many residents, Carol had strong likes and dislikes. She was partial to powder blue and insisted on that color for the room. It matched many other things she owned and pleased her, made her feel good.

Right from the beginning Carol had had one strong request—some-place to hang her clothes. After being hospitalized, she was rebuilding her wardrobe. Having her own things again was important to her.

To solve the problem of a closet, we looked for something that would fit in with the character of the room and be flexible. Finding a place to put wardrobes was difficult because three of the walls had a door or doors and the fourth had windows. There was a fireplace, as well, to add to the difficulty. The only solution seemed to be to build two small closet units in the corners, next to the windows. The two closets with the curtain rods that ran between them had the effect of framing the windows (fig. 4-92). The closet-and-curtain system was simple but looked formal and worked well with the character of the room.

With the closets in place we began to experiment with the rest of the furniture. Carol would live with one arrangement for a few days and then suggest a new one. Living with each of several setups, she became aware of preferences and determined which were more suited to her activities in the room.

One of the final things we did was to make new curtains for the windows. Her windows looked out onto a busy street, and Carol was concerned about visual privacy. We made some plain café curtains with inexpensive, unbleached muslin. The effect of light filtering through was quite rich.

Proceeding one step at a time worked for Carol. Each decision led to the next. Gradually she began to make adjustments and rearrange-ments on her own. When she had a few of the other residents in to chat, she realized that there was no place for people to rest their cof-fee cups. So she added a small table. She liked an arm chair with a red and white slipcover that was stored at the house, and soon that joined the other furnishings in the room.

For Carol, the process of seeing her room change and develop in many ways paralleled her own growth and increasing confidence. Mak-ing herself cozy in her room and keeping her lists, Carol pulled herself together fast. She convinced her doctor to reduce her medication, was more and more available to people in the house, and became a mem-ber of the committee to work on ideas for changes to the shared spaces downstairs.

4-80. Carol needed closet space, and these two small units worked well in a room with no clear wall space.

4-81. Carol rearranged her room several times; this was one layout.

4-82. This was another of Carol's arrangements.

CHANGING SHARED SPACES

Properly planned and enacted, the process of change benefits both staff and residents. For residents, self-esteem and feelings of ownership are enhanced through decision making, while for staff, the process can serve to remotivate worn-out members. The experience of one man, working in a group home for retarded women, reveals the need for such periodic morale raisers.

"Every day, a lot of my time is spent making up things for residents to do," he said in an interview, "while the real things—the things that actually make a difference about whether the house keeps functioning—are done by staff. So we focus on piddling little things like being properly dressed in public. A big deal is made out of that; so much so that we set standards for our residents that we ourselves don't follow. It's hypocritical. We always say, 'Well, we have to make them dress better or they won't be accepted.' But I think it's really because we're at a loss for things to do. We don't trust their abilities to become capable of handling total meal preparation or doing the wash on their own, beginning to end—so all that's left is this piddling little adherence to rules.

"I'd rather be working together on real things where there's some mutual respect. I know whenever I make up something for them to do I'm treating them like incompetent children. It wears me out because I know I'm faking it. They know it, too. How long can you keep up your spirits when you know most of what you're doing is fake?"

This statement typifies many staff problems. There are two parts to what this man is saying. Though he does not want to, he is treating residents as incompetent children. He has lost his spirit and no longer believes in his work—he is burned-out. Even the best, most motivated person working in a human service group can eventually reach a point at which everything and everyone seems routine. There is no energy, and it is easy to feel just plain tired of it all. Such burnout also destroys the overall effectiveness of an organization. Promises are made that are not kept; stories are made up to impress outsiders while everyone knows they are not true; staff abuse of sick time becomes a problem. Everyone feels in a rut.

One way to combat staff burnout is by changing the place. We have seen some of these attitudes turn around as staff people become involved with residents in changing their place. This makes sense, for it means breaking the routine, getting involved in something new, and working with others—both staff and residents—in a new way. When decisions are made and work is done by a group of peers, self-perceptions and perceptions of each other change; roles are less important. The feeling of camaraderie generates enthusiasm and respect; it is one step toward combating burnout.

With more responsibility, many residents become more competent. This is a normalization principle observed in many successful group homes. When residents take part in making decisions about how each room will be used, where major appliances will go, what colors walls will be painted, who has access to a space, and so on, they gain in self-esteem and competence. In too many group homes these decisions—and the action that results—are done "for" residents by staff, acting as a reminder to them of their lack of skills.

Other patterns can change too. In one group home we found that residents typically reported having a close friend who lived somewhere else and was rarely seen; there was little friendship among residents within the home. After the residents were involved in changing their own space, they became friendlier with each other and spent time together outside the house. The shared experience had brought them together.

Another pattern that changed was that fewer people hung around the dining room waiting idly before meals. This behavior is common in institutions, not only before meals but before the weekly movie, afternoon shopping trip, or other group activity. Instead, once the residents had been involved in changing their spaces, they seemed to find better things to do with their time than wait idly.

Making changes also generates new feelings of ownership. Whether in the private bedrooms or the shared living room, anyone who has a hand in making changes will feel more in control of place and self. One happy side effect of having a stake in things is that the home is usually better cared for, kept cleaner and tidier.

The process will vary with specific people and places. What is constant from one place to another is the power of the experience. In changing the place, the people within also change.

While every house will handle the change-making process slightly differently, we have found that several procedures work well to get the projects underway: establishing a design group, analyzing the existing arrangements, and mapping the usage patterns. By the completion of these basic preliminaries, the design group should have found the center of the house and formed a fairly clear idea of what changes will be beneficial. They can make their plans and conclusions known to the rest of the house in a group meeting, which might include both a slide show and discussion period. Other residents can involve themselves in the actual work phase of changing shared spaces.

The Design Group

To begin a change project, form a design group of three to five people. The group can be an opportunity for staff and residents to work together as peers. The group should then select one member to act as organizer or coordinator, preferably someone with a high energy level who can spark the enthusiasm of others. Participation may be increased if leadership is rotated occasionally. The tasks of this group will be to identify problems within the shared spaces, propose alternative solutions, and implement a small first project that best solves one problem.

From the beginning the group will need some resources including a modest budget—$50 perhaps. Large sheets of paper and markers are useful—tack sheets to the wall and begin generating lists of spaces and problems and of resources and ideas. Other needs include writing pads, sketch pads, pencils, and so on; access to a typewriter and copy machine will also prove valuable. If need be, borrow a camera to take slides both of the house and of other places and homes to illustrate problems as well as desirable and attractive spaces.

The organizer should assure that every meeting has a real purpose. Enthusiasm will die quickly if meetings are called without adequate preparation. At the end of the meeting, design group members should feel that something was accomplished.

Some meetings are for discussing plans; others are set up to actually get something done. Do not let meetings be just all talk. The inclusion of some action will keep up the group's enthusiasm.

The primary concern ought to be that each person in the house has a chance to participate—if not in the design group, then certainly in the implementation of change. Do not forge ahead, no matter what, just to get the job done. The experience of being involved in the change process is as important as the actual change itself. Staff members may have to give up some of their control to let the group find its own way.

After the design group has agreed upon and prioritized problems, it is time to gather the entire household for a group meeting. Do not move on to solutions before giving others a chance to have their say. Use the slides you have taken. They are a wonderful tool for stimulating discussion. Once the problems have been redefined by the whole group and a consensus has been reached, the design group can then proceed with generating solutions, preferably two or three for each problem, which will then be presented to the whole group at another meeting.

Space Analysis

The first task of the design group will be to analyze the whole house, defining its good and bad points. It is especially important to define problems before considering solutions. One essential task of the organizer will be to keep the group focused on problem definition in the beginning. While it is natural to want to jump ahead to the solutions, this is not the most effective way to decide on changes.

In making the analysis, consider the house as a total place rather than as isolated rooms. How well do the rooms work together? Spend time looking at how the whole house works and then identify areas needing attention. To do this, there are several procedures that we have found useful that can be adapted to fit specific houses—inventories, floor plans, and behavioral mapping.

Taking Inventory

An inventory can facilitate taking stock and isolating problems. The sample in figure 5-1 can be adapted, and information should be recorded room by room. Include offices, hallways, porches, storage, and other "nonrooms."

What variety of activities can be supported in the house? Describe the range of group sizes and show how rooms can be used in more than one way. Flexibility is a key.

Drawing Floor Plans

Drawing a floor plan is really quite simple. Many people are unnecessarily critical of their own first attempts, and the best advice we can give is that it is worth doing. It becomes easier as familiarity with the house and process increases.

The first step is to sketch the plan of a single room, as if the top of the room has been sliced off and you are looking down from above. Use

plain paper and draw the outline of the room, including all doors, windows, corners of walls, niches, and radiators. This does not have to be 100 percent accurate, though it should be carefully done, for on this working sketch the actual dimensions will be recorded.

To record dimensions, enlist the aid of two other people and use a measuring tape, preferably 16 feet long. To find the dimensions, start measuring by having one person hold the end of the tape in a corner of the room. A second person unrolls the tape along one wall, stopping at each window opening, doorway, niche, or radiator and calling out the dimension until reaching the opposite corner. The third person records these dimensions on the sketch. Repeat for each of the walls of the room.

Use these sketch plans and dimensions to draw accurate floor plans of each room, this time on graph paper as shown in figures 5-2 and 5-3. If graph paper with four squares to the inch is used, let one square on the paper equal one foot in the room. The plan will then be at a scale of $\frac{1}{4}$ inch = 1 foot.

With several copies of the plan, various room arrangements and schemes can be sketched. Reduce the floor plan on a copy machine if you intend to use it for behavioral mapping. Draw in the existing furniture arrangement on one copy to use for later comparisons.

A floor plan should be made for all the public or shared areas of the house. By taping together plans of individual rooms, you can construct a

```
SPACE INVENTORY
----------------

Room: Den          Date: April 17th  By: David C
Size: 9 x 12  Traffic ( ) does
                      (X) does not  cut through.
Describe connections to other spaces: Large arch
into entry hall. sliding doors can be closed
Describe openings to outdoors: Big front window
overlooks street. side windows view yard
Describe noise problems: It's pretty good,
though traffic can be noisy
Rate condition:
1 = Excellent, 2 = OK, 3 = Needs Attention, 4 = Terrible.

2 Walls      2 Ceiling     2 Windows     2 Doors
2 Floors     2 Trim        2 Furniture   3 Odors
4 Carpet     3 Ventilation 1 Cleanliness 1 Heat

Compared to other spaces in our house, this space is:

                  Cheery (X) ( ) ( ) Drab
          Overcrowded ( ) (X) ( ) Underused

               Public ( ) ( ) (X) Private
           Frontstage ( ) ( ) (X) Backstage

                 Soul (X) ( ) ( ) Prim
                 Hard ( ) ( ) (X) Soft

Describe activities that happen here through the day.
Which happen at the same time? Reading, visiting, a
good round of clues. several happen at once
Describe signs of ownership. Who feels ownership? Why?
Books and shawl belong to any. everyone helped decorate.
Evaluate use of space, considering size and qualities.
It works well. more hobbies could happen but
they'd need better lighting!
```

5-1. A space inventory allows you to isolate problems.

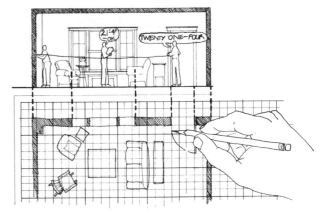

5-2. A rough floor plan can show all the elements in a room.

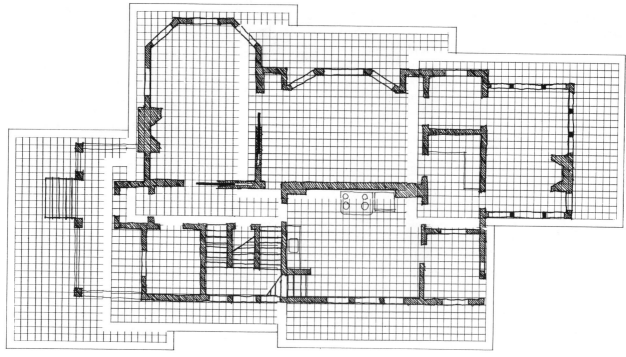

5-3. An accurate floor plan, done to scale, shows all built-in elements in a space.

plan for the whole house. Use the overall dimension of the house, front to back and side to side, to check the accuracy of your taped plan. Remember to add the thickness of the walls between rooms.

Finding the Center

Does the house have a center? It is probably no longer literally the hearth around which the family gathered: the fireplace has long since been replaced by other sources of heat. But the house will have a place that seems to be its center.

The center may shift during the course of the day as the rhythm of activity changes. It may also change through the year, as rituals and special times are marked.

The center is a comfortable place in which people linger and are part of the life of the house. It is a place where it is easy to keep abreast of things, of people's comings and goings. Here

the many divergent people and personalities of the group come together.

Where do people in the house gather spontaneously? Where is the first place to go to look for someone or ask for information? Where do people gravitate to when coming in from the outside? Chances are that everyone in your group will answer these questions the same way.

Sometimes the center is the kitchen with its liveliness, bustle, and good aromas. Other times it is the cozy seating area by the fireplace, or it might be the front hall with its views into adjacent rooms.

In one group home, formerly a farmhouse, the large kitchen is by far the most active space. The direct view of the driveway and back entry make it a natural place for watching people come and go. There is a bulletin board, large table, and constant cooking activity. Because the front rooms are slightly formal, the casual atmosphere of the kitchen makes a welcome contrast. We were struck by the warmth, coziness, and

5-4. This central space can be used for dining.

5-5. The same space shown in figure 5-4 used as a social center.

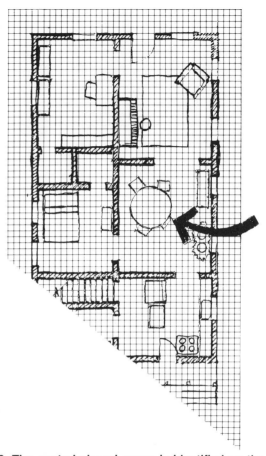

5-6. The central shared space is identified on the floor plan.

relaxed atmosphere as people drifted in and out throughout the day. There was a nice bustle that was neither distracting nor annoying.

One reason to locate the center of the home is to decide what would make it more attractive. A good center is big enough for about half the group at any one time. Extra chairs can be brought in for that occasional meeting. It should be a place that people pass by or through regularly. The main attraction of the room should not be the television set. The symbolism, warmth, and imagery of a fireplace is one kind of attraction. Feeling "in the middle of things," in the center of activity, is another attractive feature. Partly it is the feeling, the pleasure derived from being together, talking, and appreciating each member of the group.

Listening to the House

Get a few people together in one room and be quiet. What's the loudest noise you can hear? Television? Stereo? Shut it off and listen again. Now what's loudest? Refrigerator motor, furnace blower, air conditioning fan? Shut that off too.

Now what? The buzz from fluorescent lights? Outdoor traffic? Keep shutting off (or shutting out) noises. Hear any new sounds? Birds or wind in the trees? Your own breathing? Can you sense the size of the room changing with a change in sound? Do you personally feel any less stress? More comfortable?

The amount of sound in the house can be controlled. Check for sound absorbers—rugs, drapes, upholstery, bulletin boards. Perhaps a door can be used to keep out unwanted noise.

Listen for positive sounds too. Conversation, perking coffee, laughter, the clatter of dishes—all are a comfortable and familiar part of life at home.

Making Rooms Do More

In institutions, people are treated as a herd, not as individuals. Participation in activities means participation with everyone else and each space is set up to reflect this all-or-nothing attitude—as if everyone does the same thing at the same time. This is expected in institutions, but it is disturbing in group homes.

Even so, time and again, in the homes we surveyed, we came across examples. Most dining rooms surveyed were set up for large meals only, unused the rest of the day. Living rooms often resembled institutional dayrooms, with chairs enough for every resident ringed around the walls, facing the empty middle or the ubiquitous television. In many cases, kitchens were the domain of the cook only, with group members excluded. An exception to this was one particularly pleasant house in which the kitchen table was the social hub. Offices are usually the exclusive domain of staff, kept locked and empty after working hours. In another exception, a staff office was kept open at night for residents' use.

The all-or-nothing attitude erodes feelings of mutual trust, personal control, and ownership. If the spaces in the house are set up for one purpose only, residents will find themselves concentrated in one room with other spaces virtually empty. The individuality of each person is diminished, as is the contribution of each to the life of the house.

In analyzing the house and planning changes,

the design group should pay special attention to the multiuse potential of the shared spaces. The process of drawing floor plans will help visualize this potential (figs. 5-7 and 5-8).

Behavioral Mapping

The most important decision to be made when choosing the first project involves determining which space must be examined and which behaviors and activities are important. Once the particular space is chosen (the whole downstairs, just the dining room, the dining room table only) and a decision is made about whose activities to observe (everyone, staff and residents but not visitors, residents only), the next step is to determine what to observe: activities, interactions, specific behaviors or broader types, simple location, body positions, or movement. It is best to start out simply; more categories can be added after some experience. Trying to record

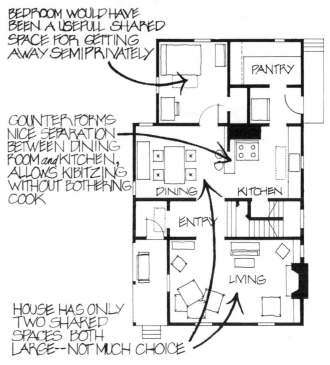

BEDROOM WOULD HAVE BEEN A USEFULL SHARED SPACE FOR GETTING AWAY SEMIPRIVATELY

COUNTER FORMS NICE SEPARATION BETWEEN DINING ROOM and KITCHEN, ALLOWS KIBITZING WITHOUT BOTHERING COOK

HOUSE HAS ONLY TWO SHARED SPACES BOTH LARGE--NOT MUCH CHOICE

PANTRY

DINING   KITCHEN

ENTRY

LIVING

5-7. Drawing a floor plan can help residents identify strengths and weaknesses of a house.

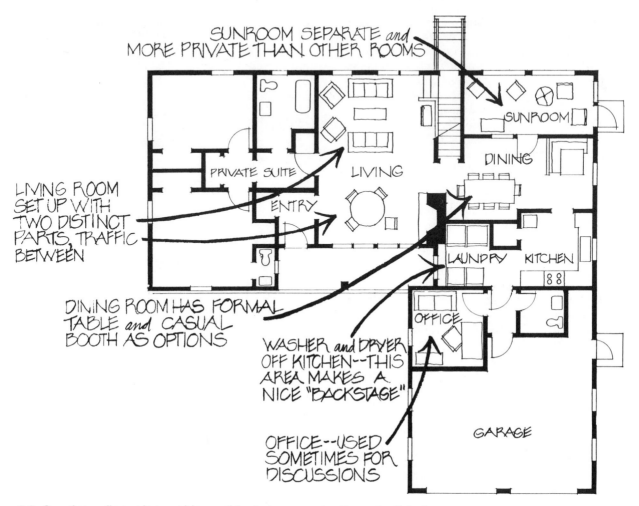

SUNROOM SEPARATE and MORE PRIVATE THAN OTHER ROOMS

SUNROOM

DINING

PRIVATE SUITE

LIVING

LIVING ROOM SET UP WITH TWO DISTINCT PARTS, TRAFFIC BETWEEN

ENTRY

DINING ROOM HAS FORMAL TABLE and CASUAL BOOTH AS OPTIONS

LAUNDRY   KITCHEN

WASHER and DRYER OFF KITCHEN--THIS AREA MAKES A NICE "BACKSTAGE"

OFFICE

OFFICE--USED SOMETIMES FOR DISCUSSIONS

GARAGE

5-8. Creating a floor plan enables residents to recognize the potential of shared spaces.

too many things at once will prove confusing.

The basis for making these decisions will be the problem selected by the design group as needing the most attention. Perhaps it is the underutilization of the dining room during non-meal hours or the lack of interaction during the meal. The problem to be resolved will define which space, activities, interactions, and behaviors will be used in behavioral mapping. A mapping of the dining room will show who is using the dining room during the course of the day and what kinds of activities are occurring there. This information will be useful in deciding how the room should change. After changes have been made to the room, a second mapping will show whether those changes have successfully resolved the problem or not.

The basic process of mapping is simple. First use a floor plan of a particular space (the dining room in this example) to record individual locations and corresponding behavior. Next, make several more maps of the room at other times of the day. Put the several maps together to form a composite of all the behaviors in that space.

To begin, use the floor plan prepared earlier and include the position of all furnishings in the room. Use a copy machine to reduce the plan to cover about half a sheet of paper. The other half of the page is then divided into columns to record each person's location, activity, interaction, or other behavior. These categories will depend on the particular circumstances and problem addressed. Under the heading for interaction, begin with "talking," "nonverbal interaction," and "no interaction." For activity, it is best to do some preliminary observation and simply write down every activity that occurs. From this preliminary list, generalizations can be made. Simple categories are "eating," "meal-related" (such as setting or clearing the table), and "other." You can also use more specific activities such as "reading," "talking," "sleeping," "staring," "no apparent activity," "playing piano." The list of activities should not be too long, or it becomes difficult to draw conclusions.

A little space on the page will be needed to mark down the name of the mapper, time of day, and date. Finally, leave space to jot down a short reminder note about unusual occurrences during the mapping (fire drills, birthday parties). Use the sample (fig. 5-9) as a guide, adapting and changing it to suit the location and circumstance.

Decide whether it is important to have data on each individual or only on the group as a whole. In some houses the keeping of individual data is considered an invasion of privacy. This ethical question should be discussed in a house meeting, and a decision should be made about the mapping process. If the decision has been made to keep data on an individual basis, it is easiest to assign a number to each person and record that on the plan, rather than to write out the name of each person. If only group data will be kept, simply make a mark for each person without identifying individuals. The entire mapping process should, of course, be thoroughly explained to all members of the household. With a full understanding of the process and its purpose, most people will go about their normal routines and not be self-conscious in their actions.

Once the blank form is prepared, actual mapping begins. The idea is to record a "snapshot" of the room. Observe the whole room and begin, as quickly as possible, to write down each person's locations, activity, interaction, until all the

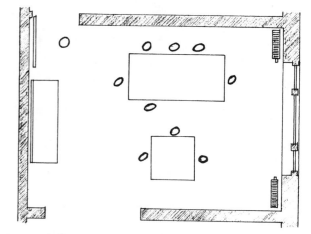

PERSON			INTER-ACTION			ACTIVITY								BODY POSITION			
resident	staff	other	talking	non-verbal	no interaction	eating	talking	sleeping	radio	cards	cleaning	meal related	no activity	sitting	standing	leaning	walking
✓			✓			✓								✓			
✓			✓				✓							✓			
✓					✓	✓											
✓					✓	✓							✓	✓			
✓					✓	✓					✓						✓
	✓		✓			✓								✓			
		✓	✓				✓							✓			

5-9. A sample behavioral mapping form.

categories for each person are filled in. The mapper must make quick judgments. Someone may be talking and eating at the same time. Are they doing one, the other, or both? People will move around and do different things while the mapper is trying to map—they will not hold still; it takes a bit of practice to get everything down.

The next decision to make in mapping behavior concerns the number and frequency of maps.

In order to get a true picture of how a room is used, it should be mapped over several days of the week and at different times of the day from morning till evening. Complete one map every fifteen minutes between 9:00 A.M. and 10:00 A.M. on Monday. On Tuesday, map between noon and 1:00 P.M. About twelve hours of mapping, distributed over the days of the week and hours of the day, should provide a good base of data from which to draw conclusions.

Next the mapper will compile all the maps to form a composite. Count the number of times on all maps that someone was recorded standing by the door, staring and not interacting. How many times were people mapped standing as opposed to sitting? Continue with all categories until the composite map is complete. This composite will show what parts of the room are most used and what activities happen most frequently. Present the findings to the design group. It is a good idea to make a composite before and after making a change in the room. Compare the maps to see how activities and behaviors have changed as a result of the group's effort.

Preparing a Slide Show

A slide show might be part of the group discussion or it might be a special event. In any case it is certain to be one of the most exciting happenings in the house. Slides are superior to snapshots—projected on the wall they are large enough for all to see and discuss. Slides can provide a fresh view of an environment taken for granted during the day-to-day routine.

There is ample opportunity for everyone in the house to be involved in making the slide show. Of course, there must be someone who can use a camera and who, in turn, can show others how to take slides so several different points of view will be represented. Part of the design group budget should be allocated for film and developing.

The camera crew should shoot a wide variety of subjects. One theme might be the signs of ownership and control throughout the house—a picture painted and hung by a resident or a staff member's flowering cactus. Another subject might be just the entrance, concentrating on the sorts of messages—welcoming or otherwise—it might give to visitors. Or take slides of neglected areas in the house or of one person's three favorite aspects of the house.

Look for details as well as for overall views. A detail shot—such as of those neat individual mail slots in the front hall—can say a lot about how attention is paid to individuality. On the other hand an overall view can capture atmosphere and can show, for instance, how harsh lighting in the living room tends to make the whole room seem cold and uninviting.

Once the slides are made, the design group should decide on the order of the show. Perhaps this will be room by room or by theme. Photographs of the messiest parts of the house can be grouped together or areas of the house in which color is used most and least pleasantly can be categorized.

One effective way of showing slides is to set two projectors side by side, to show two views simultaneously of the same area or to compare the living room with the living room in another house.

The people who took the photographs should have a say in how they are organized. Try to limit the total number of slides to about thirty-six, or the audience may understandably become restless.

Before the slide show is held, the design group should make some basic decisions about its format. Will everyone respond to a question-and-answer approach, or would something more playful be effective? For example, a prize might be offered for the person who can spot the greatest number of homey touches during the show. Or one person might explain a slide of one room to the rest of the group, as if everyone were visiting the room for the first time.

Once all the preparations have been made, the slide show is something to be enjoyed. While it might be useful to have someone taking notes, we suggest that for the most part everyone simply enjoy the moment for its own sake. One of the nice things about working to change the place is the sharing of perceptions, the common goals, and the slide show gives everyone something real to talk about.

Group Discussion

Before the enjoyment of the slide show wears off, introduce the shared space projects that follow in this section. Review the slides for consensus on the most pressing problems and discuss shared space projects in relation to these problems. Once there is agreement on general directions, it becomes the task of the design group to begin proposing projects to the group.

Moving the Television Set

As a first project, moving the television set produces surprising results. There is a wide range of opinions about television. Some feel it is a valuable connection to the outside world, while others see it as an excuse to sit and stare and therefore view it as a dangerous drug, thinking "the less the better." However, regardless of attitude, most group homes have allowed television to occupy center stage. It is the dominant feature of shared spaces, often dictating the arrangement of the furniture. Chairs set in theaterlike rows discourage discourse among viewers.

5-10. A group discussion among all the residents is a vital part of the redesign process.

At center stage, the set will be left on whether anyone is watching or not. Suddenly another day of interminable television-watching has passed. Options are effectively reduced; choice-making is discouraged. The din invades the rest of the house—kitchen, dining room, and library—drawing people to itself almost hypnotically.

A good first project for the design group is quite simply relocating the television set. Television has a lot of power in our lives; taking control over it is a major assertion. In one shared-space project we worked on, it was hard to get a reaction until we talked of moving the television. There were as many opinions as people in the house and possible places to put the set. The debate was a lively one, with everyone in the house involved.

Issues and problems were openly discussed. Disadvantages of the present location were determined by asking whether television viewing interfered with visiting, conversation, and other activities, and whether people sat passively watching. How much noise traveled through the house was another topic. The positive effects of the new location were also considered—more active participation in a wider range of house activities and less aggravation.

The new location for the television should not be one that is already popular or the choices will remain as they were. However, if you move it to an area that is not much used, options will increase. You will have the original location that is already popular as well as an additional area that uses television as a magnet.

The best locations are away from the center of the house, preferably in a space that can be closed off. A small alcove off a large room works well, or a small staff office may be used infrequently enough to make a good television room. Developing basement space is made easy by moving in the television. The set draws people to it, and its noise is compatible with Ping Pong, pool, or other such activities.

Another choice is to combine television viewing with some other activity, such as doing the laundry. This brings some life to clothes washing, because there will usually be someone watching television who will keep the launderers or ironers company. Television itself is less boring if viewers can also do something with their

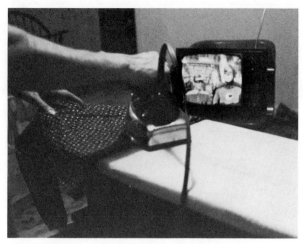

5-11. The television can be moved to the laundry area, where watching can be combined with other tasks.

hands. Both activities benefit from being combined.

Whatever new viewing area is decided upon, it should be arranged to accommodate a variety of television viewing styles. Instead of rows of chairs, a table and chairs allow for easy interactions among viewers and also encourage other activities concurrent with watching—card playing, clothes folding, or letter writing. Fat pillows on the floor accommodate quite a different watching style—casual, cozy, and fun. A couple of comfortable chairs add another choice.

Before making the big move, the design group should canvas the entire house for preferences. As obvious as this sounds, it is sometimes easy to forget. This can be done informally by simply asking, or formally by making a checklist similar to the sample shown in figure 5-12. The value of checklists is the thought required in making one. Thinking must be focused so that issues are distilled to those few important items that will constitute the checklist.

Checklists should be completed for at least two different locations—the original site (the den in our example) and the planned one—in order to compare before and after. A checklist cannot determine whether a location is in itself good or bad—only whether one is better or worse than another. For each location, fill out at least six copies of the checklist. Taking only one sample

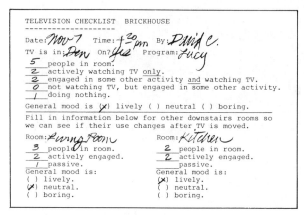

```
TELEVISION CHECKLIST  BRICKHOUSE
---------------------
Date: Nov 7   Time: 7:20 pm  By: David C.
TV is in: Den  On? Yes  Program: Lucy
  5  people in room.
  2  actively watching TV only.
  2  engaged in some other activity and watching TV.
  0  not watching TV, but engaged in some other activity.
  1  doing nothing.
General mood is (x) lively ( ) neutral ( ) boring.

Fill in information below for other downstairs rooms so
we can see if their use changes after TV is moved.

Room: Living Room        Room: Kitchen
  3  people in room.        2  people in room.
  2  actively engaged.      2  actively engaged.
  1  passive.                  passive.
General mood is:          General mood is:
( ) lively.              (x) lively.
(x) neutral.             ( ) neutral.
( ) boring.              ( ) boring.
```

5-12. A checklist of design alternatives requires residents to think about their choices.

will not assure results that are truly reflective of the situation. For a fair comparison, checklists for each site should be filled out on the same day at the same time.

The issues discussed earlier can form the basis of the checklists. They should be kept simple, with items to count or check off and questions to answer with "yes" or "no." Rooms other than the one the television is in should also be evaluated.

In one house, we used behavioral mapping and a short questionnaire to evaluate the success of the move. The television had been taken out of the living room and moved into what had been a practically vacant space. The results showed changes in activities and the use of space. Before the move, the living room had the highest concentration of use, while connected social spaces —music room and sun room—were little used. After the move, use of the living room declined (39 percent to 18 percent) but use of the three rooms as a whole remained roughly the same. Concurrently, game playing as an activity increased (1 percent to 20 percent), primarily in the music room. In other words, activities shifted from passive television viewing to more active ones, in smaller, dispersed groups, which made better use of the available space.

Residents' questionnaire responses were favorable too. With the television moved out, the living room was still considered the center of activity (58 percent before, 60 percent after). While 75 percent of the residents had considered

the living room the noisiest in the house before, only 50 percent thought so after. When asked which room sometimes had too many people in it, 58 percent of the residents answered "living room" before the move compared to 36 percent after. It seemed less crowded.

Although people were still watching about as much television as before, they now considered the living room less noisy and crowded, but still the center of house activity. People were now making a choice to watch, to take advantage of other space in the house, and to engage in a wider variety of activity. The experiment was a success!

Rearranging the Living Room

Try to visualize a few people doing something with no furniture or props at all before actually arranging furniture. It will take a little practice at first to get a picture in your mind. Start with a familiar situation, for example, four people playing cards. The people without furniture and props will be sitting within touching range of each other, about 4 or 5 feet apart, face to face. Their body postures are upright—more than if they were in easy chairs but less than at a formal dinner. Their knees are quite close together, much closer than if they were sitting in the open, talking.

Once the setting is visualized the props can be imagined—cards, a score pad, a few drinks, and a bowl of popcorn. Furniture will include a table and chairs. A hanging light overhead helps tie the setting all together. A nearby window allows an occasional view outside while a doorway gives a glimpse of another group working on a project. Circulation space allows access to the chairs.

Obviously, this setting is a social one. It is larger than a single piece of furniture but not usually as large as a whole room. Settings range in size, accommodating one person in a quiet reading chair next to the fireplace or the whole group around the dining room table. Settings suggest and support activity; a single setting may support many different activities at different times. In the example given, the card table may also be used for eating, doing a jigsaw puzzle, or writing letters.

Almost every room can accommodate several settings for different activities and group sizes. The room does more; it invites participation and activity through the day.

Simply arranging furniture will define supportive settings. Furniture is meant to be moved, and residents should feel comfortable doing so. Arrangements can be tried and rearranged until the group is satisfied. A floor plan is a helpful tool for looking at several arrangements. Furniture cutouts can be made to scale and moved around the plan. Still, there is no substitute for the real thing; it is easy enough to put everything back again.

The room itself may suggest an arrangement. Some special aspect of the room—a cozy, quiet corner, a window with a view, a windowseat, or a fireplace—is a good place to begin. It can be arranged with a specific activity and group size in mind. A freestanding screen will make it more flexible.

A variety of furnishings should be available in the room, besides the usual chair, couch, and coffee table. A writing desk, wicker peacock chair, loveseat, rocker, drafting table, footstool, open or glassed bookcase, giant stuffed pillow, or roll-top desk make good additions. A range of old and new, light and heavy, sleek and overstuffed furniture is best.

Creating supportive settings often gets the furniture away from walls and out in the round. However, people can be uncomfortable, sitting out in the middle of a space with their backs exposed. A heavy couch with a deep back helps; putting a writing desk, library table, or low bookcase behind the couch will make it seem even more protected. Lighting and such details as area rugs can also anchor a setting.

In one house, we went a step beyond floor plans and built a model to bring our ideas to life.

Building a model gave the design group an opportunity to work, making it easy to visualize changes in the real space by seeing them in a model. Moving a couch or piano is simple enough in the model, making it likely that many possibilities will be tried.

We also continued experimenting with the room itself. By trying each arrangement for a few days, everyone had an opportunity to use it and react.

To involve other residents and staff, the design group used simple questionnaires and checklists. There were two main issues. What does the group do in the living room now? What would the group like to be doing in the living room? There were both lists of activities to check off and open-ended questions to encourage response.

Among the responses were suggestions to make the living room a pinball or pool room, a snack bar, or an exercise room. These ideas pointed out the need for an active area somewhere in the house, though they were finally incorporated into other rooms. Consider a full

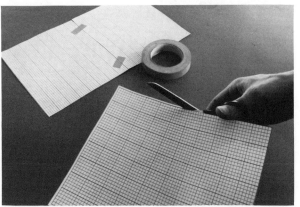

5-13. To build a model, first trim graph paper and tape it together to fit the shape of the house.

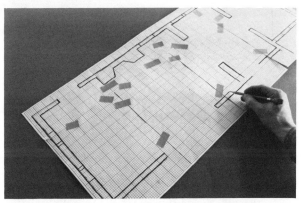

5-14. Transfer the floor plan to the graph paper.

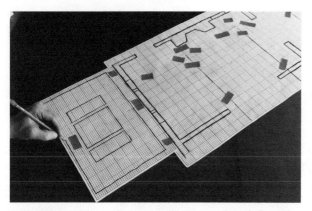

5-15. Now add graph paper to one end and draw in the elements of that endwall.

5-18. Using the drawing as a template, cut out the wall, including window and door openings.

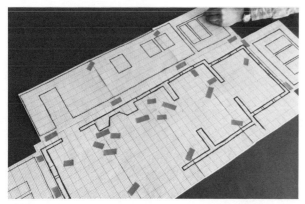

5-16. Draw the other three walls in the same way.

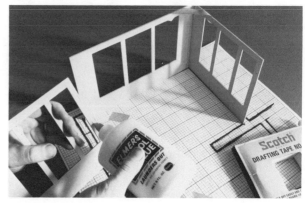

5-19. After making each of the cardboard walls, glue them to the floor plan.

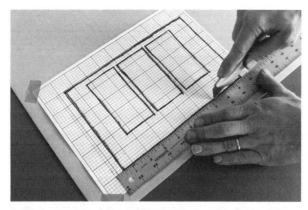

5-17. Lay one of the wall plans over 1/16-inch chipboard or cardboard.

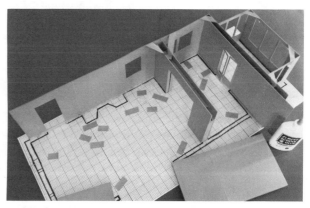

5-20. Remember to leave appropriate space between walls to approximate their thickness.

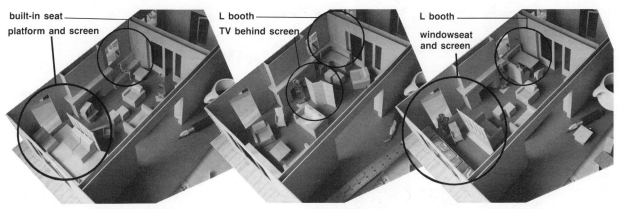

built-in seat
platform and screen

L booth
TV behind screen

L booth
windowseat
and screen

5-21. Build furniture to scale as well; it can be rearranged within the model to illustrate the design alternatives.

range of ideas, even unrealistic ones, before setting limits.

Focusing on a specific problem—the need, say, for a quiet reading spot—is another way of generating ideas. For many people, concentrating on a single part of a room is not as overwhelming as tackling the whole.

Along with the design group, a "refurbishing crew" was organized, providing an opportunity for others to participate. While the planning continued, refurbishing of the living room began. This included a general cleaning, painting, and making necessary repairs. For those who did not want to plan, this was another way to be involved.

The design group set up three living room plans in the model. Three is a good number from which to make a choice. Even after changes were begun, the plan was modified. When the L-shaped booth arrived, it was put in place and somehow was not quite right. After living with it for a few days, someone suggested turning it 90 degrees, which was better, but still not right. After a time, it was moved to the music room where it fit perfectly and was an immediate success, being in almost constant use (fig. 5-23).

Trying out several plans in the living room or any shared space is the best way to get people to participate. Having different plans on the table opens the way for opinion making and discussion. In the shared-space project described above, we found a definite increase in camaraderie at the house, perhaps the most dramatic

5-22. The final living room design of the house shown in the model.

5-23. The final music room, adjacent to the living room, of the house in the model.

5-24. The final sun room, adjacent to the music room, of the house in the model.

result of participation in an environmental change we witnessed. Camaraderie—that special friendship that comes in willingly sharing a space—is just the quality that makes life in a group home different from either the stifling authoritarianism of an institution or from the isolation of life alone in an apartment. We were convinced that the feeling had increased through participation in the design project.

The residents in the house felt better and more comfortable about living and being there. When asked whether they liked living in the house, 71 percent of the residents said yes before the project, 93 percent after. When asked whether they liked being with other residents, 64 percent said yes before, 100 percent after. While most people did answer yes to having a special friend, only 17 percent said the friend also lived in the group home. After participating in the project, 44 percent considered a fellow resident a special friend.

These significant gains in the closeness of the group were also evidenced by changes in activity. Cooperative activities increased while solitary activities decreased. Game playing as a preferred activity jumped from 1 percent to 20 percent of the activities we included in behavioral maps. People were using their leisure time in more socially interactive ways than before the project began.

Changing the Dining Room

The group meal—the one everyone is expected to attend at least once a day—is a goal in most group homes. It is a symbol of togetherness. The idea is to share the day's experiences, to enjoy friendly talk, and to reinforce a sense of being part of the group. Typically, however, this symbolic group meal falls short of expectations. In our work in group homes we made it a point to attend this special meal and, unfortunately, have found it usually to be an eat-and-run affair.

Our impression is that while it takes a real effort to prepare the meal, very little attention is paid to how the group gets together to eat it. Roles are often rigidly defined—staff as the providers and residents as the passive consumers. For the group meal to symbolize the whole group, all participants, residents and staff alike, must be part of the entire meal ritual—preparing, consuming, and cleaning up. There is a spirit of mutual respect—which in turn lets conversations flow—in a group where everyone takes part.

The dining room, the site of this important ritual, must reflect the special style of the group. We have also observed that if conversation is lively, chances are good that the meal is a success. Of course, dining room arrangement does not guarantee lively conversation, but the shape of the table does help, just as the character of the room itself does. An intimate restaurant will encourage diners to linger and talk while many a cafeteria encourages people to eat quickly and make room for the next customer.

Some groups may benefit by having several small tables; a single large table may fit others. And not just any table will do but one that suits the house and group—consider refinishing an old oak table, for example. The table selected will obviously depend on the size and shape of the dining room, the group size, and mealtime style.

Round tables put everyone on an equal footing, for there is no head of the table. It is as easy to talk to the person seated next to you as to those across the table. One disadvantage is that several small round tables cannot be grouped together to form a larger unit.

5-25. The dining room should be a space for lively interaction.

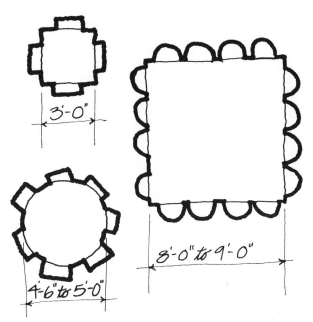

5-26. Large and small squares or circular tables promote conversation.

Square and rectangular tables have other things to offer. Corners for example are good for conversation. Sitting across a corner is most conducive to interaction—it is not as formal as looking directly across the table. This suggests that square, not rectangular, tables might be best; however, a small one accommodates only four people and may be impractical. Tables between $2\frac{1}{2}$ and 5 feet across encourage easy, comfortable talk. Square or rectangular tables are also easy to arrange and group together to form a larger surface.

Long skinny tables are rarely successful. Much over 5 feet long, they tend to line people up in straight rows, inhibiting interaction. Even with just three people in a row, the middle person must face away from one person to talk to the other. Usually the table breaks into subgroups at the opposite ends.

Long tables may be appropriate and impressive in board rooms, but an authority figure at the head of the table is out of place in egalitarian living. Also a long table will usually fit into the room one way only, limiting options and choices for a variety of arrangements.

The room itself should not be restricted to mealtime use only. Just as it is all right sometimes to eat meals in the kitchen, it is also all right to do many things in the dining room. A small couch, rocker, or piano will increase its possibilities, and small meetings can be held there. People will be attracted to what had been

5-27. Long, narrow tables limit the arrangement of other furnishings.

a dead, underused space, and the other rooms will be less heavily used.

At one house, the design group started with four existing square tables and tried arrangements that kept the four tables separate. They also tried grouping them in pairs (fig. 5-28). Each arrangement was left for a few days for reactions.

Finally one resident suggested putting the four tables together to make one big square (fig. 5-29). Although this seemed contrary to the notion that smaller groups were more conducive to interaction, the big table was an immediate success.

Other than the one weekly house meeting, meals were the only time the whole group got together. The one large table seemed symbolic of their togetherness. There appeared to be more affection expressed for it and for using it, with people lingering after the meal to chat over coffee.

One explanation for this success may be that the table created more options for use. You could be part of the group without the overbearing intimacy that small tables may force. People easily positioned themselves near to or far from others. Each corner supported its own conversation, yet it was easy to join in from across the table.

The table reinforced the connection between each person and the group as a whole. It represented camaraderie. Indeed, after this table had been in use, residents more often said that their special friend was someone living in the house rather than someone living outside. The rituals of mealtime carry over to other aspects of life in the home as well.

5-29. Some groups find one large table conducive to camaraderie.

Enhancing Nooks

People tend to think of group homes in terms of basic rooms: living, dining, bed, and bathrooms. This kind of thinking is fairly abstract and does not consider the personal experience of day-to-day use. The parts of rooms, such as nooks, give texture and define use. It is this quality that is overlooked and undervalued.

Nooks have real value. They increase the sense of control and ownership for residents and staff. Because they are small and have clear boundaries, they can easily be taken over by one or two people or by a small group seeking privacy but not seclusion. This ownership may pass from one group to another during the day. A nook may be seen as a "secondary" territory, unlike a bedroom, which, as a primary territory, must belong exclusively to one or two people.

Because the amount of furniture in a nook is limited, it can be easily arranged and rearranged for one or four people. Because the group is small, each person can affect the conversation and can be heard. These qualities make people comfortable and at ease.

Nooks increase the range of activities and social experience. More than most other spaces, they invite a range of nonspecific activities, from eating breakfast to playing cards to having a heart-to-heart talk. Nooks accommodate clothes folding, coffee breaks, letter writing, crying,

5-28. Pairs of small square tables may promote interaction.

5-30. Small groups seeking privacy can find it in a nook.

5-31. Booths can be used as cozy nooks.

laughing, and reading the morning paper. They can be a "safe" place for a new person coming into the house; it is all right to sit there and watch the action from a slight distance.

If the house already has a nook, it can be made more personal, more inviting, by adding a tackboard or ledge and by paying attention to lighting.

Homasote is an excellent material for a tackboard. It can be installed directly to a wall with panel adhesive or finishing nails, and a frame is not necessary. The tackboard should be large, maybe even wrapped around the whole inside of the nook, to encourage residents to use it for pinning up snapshots of birthday parties, notes and notices, things to do, places to go, and anything else.

A very narrow shelf or ledge, just 4 to 6 inches wide, can hold knickknacks, trophies, and paperback books.

As for lighting, ceiling lights should be avoided for they invariably cast harsh shadows on faces. Fixtures with exposed bulbs that glare should also be avoided. A hanging fixture, just a bit above eye level, works well. It neither glares nor casts heavy shadows.

If the house is without a nook, one can be created. In an old house, a butler's pantry between the kitchen and dining room can be converted easily. The location works well, of course, for a

5-32. A platformed nook is set off from, but not closed to, other spaces.

breakfast nook. The conversion may require the removal of built-in cabinets or the like and the addition of such furniture as a restaurant booth with seating (fig. 5-34).

If the house has no space so easily converted, a nook can be created in part of a larger room, preferably by adding one straight wall in a corner area, an appropriate distance from the existing parallel wall. The area should be roughly square, about 6 feet on a side—or a little less—and close enough to some activity so that it is easy to keep tabs on whatever is going on there (fig. 5-35). A view to the kitchen is terrific; so is a view to the living room or the most used entry

where the interesting comings and goings actually happen. Try to find an area with a window—the view and fresh air are both welcome.

Booth seating in either an L-shape, U-shape, or an ll-shape, with a more or less square table about 3 feet across, helps define the nook and makes it easy to slide in and out and to adjust social distance. Keep both ends open so that it is less likely that anyone will feel "trapped." A 3-foot table is narrow enough to allow close conversations, without speakers feeling too close for comfort or overly vulnerable. A table of this size also functions well as an eating or work surface and leaves space underneath for legs and knees.

Some houses have raised the seating and table onto a 6-inch high platform (fig. 5-36). The raised platform lets those sitting in the booth be close to the eye level of someone standing, so that the standing person is not intimidating those seated by looking down on them. It is for this same reason that barstools are raised about 6 inches above usual chair height, making it comfortable for someone standing to talk with someone sitting on a barstool—or in a raised booth.

The newly made nook should be mostly enclosed on three sides, like a cave, with approach more or less limited to one direction (fig. 5-37). This creates a sense of security yet

5-33. Nooks that have many uses can be found all over the house.

Booth seen from kitchen—good place to be near an activity

Same booth has view of living room and front entry

5-34. A cozy nook can be created from a pantry or other small space.

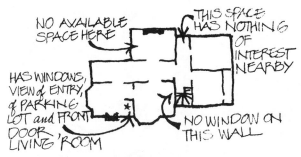

5-35. A nook can be an area partitioned from a larger room.

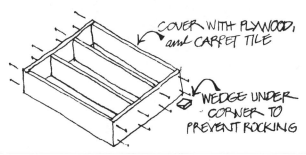

5-36. Platforms can be built to accommodate nooks.

5-37. A nook should ideally be enclosed on three sides.

with a connection to the surroundings. It is this combination of connection and enclosure that makes the nook work as a secondary territory, where ownership may pass from one person or group to another through the course of the day.

In many houses, there will be objections to breaking up any of the large spaces. Many people seem to feel that the shared spaces must all be large enough to accommodate the entire group at once. We prefer to keep one space large enough for meetings and to separate the other spaces into smaller, flexible parts. The result is that there are more places and choices for residents in the house. Making smaller, differentiated parts within a larger space does not destroy the larger space. Done with care and an eye toward flexible, multiple use, such an alteration retains the expansiveness of the larger space.

Window Seats

Window areas make especially good nooks (fig. 5-38). If the house has such a window nook, of course, it makes a special area for plants—an area where they will probably be appreciated and cared for. A board with pegs from which to hang plants is easy to install across the top of a window frame (fig. 5-39).

Even without a nook, dressing up a window area can be a great morale booster. A good window washing session may be the place to start; as dull as it sounds it can be a way of getting everyone to start talking about what might be done to the windows. Would drapes, shades, thin airy curtains, or venetian blinds be best for soft-

5-38. Window seats can be comfortable nooks.

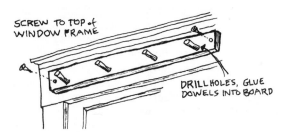

5-39. A board with pegs for plants can be mounted above a window frame.

ening the light and giving privacy? Libraries often offer do-it-yourself booklets on window coverings, which may inspire an unusual treatment.

A Note on Furnishings

Identical furnishings throughout give a group home an institutional feel; a variety of furnishings is normal and homey. Add pieces one at a time as needed—residents from the design group can do the choosing. While a total hodgepodge is to be avoided, not everything has to match. Used furniture can be a bargain, and refinishing it is a good house project. The quality of older pieces is often superior to newer mass-produced furniture. Custom-fitted slipcovers can transform an old couch or chair. A rocking chair is always a welcome addition. A ledge and tackboard (fig. 5-40) is an easy way to encourage residents to contribute and feel at home. Creative decorating will be almost immediate.

While durability is a consideration, not everything has to be indestructible. Given a chance, most people will treat delicate things delicately.

Lights

Lighting has much more effect on how we perceive and use space than many people realize. A warm glow attracts us; a harsh glare or a bland light irritates. Lighting gives cues for activity—

5-40. A ledge and tackboard make a convenient message center and encourage contribution.

an area for lively games is lit differently from a nook for quiet talk or reading.

Define the activities important to the group and emphasize a lighted place for each. No room should depend on a single light source. Instead, each part should have its own pool of light.

5-41. Area rugs help to define spaces.

Rugs

Area rugs outline spaces and underline the variety of group sizes in which residents feel comfortable. A rug can strengthen and tie together a furniture grouping. With wall-to-wall carpeting, a contrasting throw rug can be laid on top. Area rugs will complement a freshly sanded floor.

Refinishing neglected wood floors is a rewarding project for a group to undertake. The color is warm and rich, making the whole space feel that way; furniture and areas are crisply outlined. Equipment can be rented from most hardware or home improvement centers.

5-42. A finished wood floor sets off a handsome shared space.

INDEX